POULTRY FANCIERS' LIBRARY

BANTAMS

A Poultry Classic Reprint

Yours truly

BANTAMS

BY

WILLIAM FLAMANK ENTWISLE,

SCIENTIFIC BREEDER OF BANTAMS, AND LATE (EDITOR POULTRY)
OF "THE BRITISH FANCIER."

ILLUSTRATED BY LUDLOW

Beech Publishing House
Station Yard
Elsted
Midhurst
West Sussx GU29 0JT

This Impression (Third Edition) 1997

ISBN 1-85736-333-7

Beech Publishing House
Station Yard
Elsted
Midhurst
West Sussx GU29 0JT

PREFACE BY THE AUTHOR.

Inquiries have of late years frequently been made for a book on Bantams, and the only answer forthcoming hitherto has been, "There is none that we know of," and the inquirers have probably been referred to "Wright's Illustrated Book of Poultry," which undoubtedly contains the most valuable chapters on Bantams yet published.

But a volume of nearly six hundred pages is too large and costly to meet the case, and the writer of the following pages has again and again been urged to undertake the task of bringing out a little treatise upon Bantams only. This he has at length done, and in offering it to the public, while conscious that it may not contain a full answer to every possible question that might be asked on the subject, he still ventures to hope that it will be a welcome and useful addition to our present poultry literature, and, in a great measure, supply a long-felt want.

PREFACE BY THE EDITRESS.

The Author of this work on Bantams had not quite completed it (although the first parts of it had passed through the columns of *The British Fancier*) when, in October, 1892, he was compelled to lay down his pen, and, after only a few days' illness, was removed from our midst. This is not the place to enlarge upon what his loss is to his family, or to the "fancy" generally—that is left to others. But, in finishing (from his notes) the last chapters, and in editing the book, his daughter has endeavoured to carry out his wishes with regard to it; and while fully aware that in many particulars it comes far short of what it would have been had the Author been spared a few weeks longer, yet she trusts that it will fulfil his object, and furnish the fullest and most reliable information on Bantams and their culture, in a convenient and not too costly form.

CALDER GROVE HOUSE,
NEAR WAKEFIELD.

Various Bantams (Painted by J W Ludlow)

PUBLISHER'S FOREWORD

William Flamank Entwisle was probably the most important figure ever involved in the Poultry Fancy. He was responsible for introducing many new bantams taking the large fowl which existed and then, by "scientific breeding" he bantamised them. Moreover, he analysed the requirements for establishing a *standard* for each breed thereby laying the foundations for the modern day Poultry Club Standards. He was truly a remarkable man.

The other contributor to the book J. W. Ludlow was one of the outstanding poultry artists of Victorian Times. He drew 'models' of the ideal birds, thus providing a pictorial representation of the standards.

This book represents an historical record of the development of the Bantam Fancy. The inclusion of management, breeding, judging, showing and the first *standards* also make it of great practical value.

CONTENTS

LIST OF ILLUSTRATIONS

PLATES

GENERAL

BANTAMS.

INTRODUCTION.

BANTAMS, we suppose, always have had, and always will have their admirers, and, we think, in increasing numbers as the years roll on; for, regarded from any standpoint one may choose, they have much that can be said in their favour.

In Bantams we have every variety of form, and every shade of colour known in the larger breeds of fowls; we have the long-legged Game and Malays, and the short-legged Pekins, Japanese and Burmese, with their bodies quite touching the ground as they shuffle along. We have feather legged, and clean legged—rose combed, strawberry combed, pea combed and single combed, and some with no combs at all. We have the crested Polish, Sultans, Burmese and Silkies; and the whisker-booted without crests; and for perfection of markings and beauty of effect, we have the grand little gold and silver Sebrights, the greatest triumph of the breeder's art; and all these recommendations are coupled with such small size, and quiet and contented dispositions, that these Bantams can thrive, and be admired and enjoyed, in places altogether too small, and totally unfitted, for keeping larger fowls in.

In these days, objections are often made by neighbours against the crowing, in the small hours of the morning, of the large birds, such as Brahmas and Cochins, etc.; but we think we have never heard of a Bantam cock being brought before the magistrates for such a grave offence. Bantams, too, are small eaters, fairly good layers, of larger eggs, in proportion to their size, than most breeds (and the most delicious eggs, too, they are), and as mothers they quite stand at the top of the list.

They are easy to manage, and with ordinary care and attention will be found the most satisfactory pets for the younger members of the family, as they readily become so tame and confiding, yet withal so cunning, that 'tis no wonder we are constantly told that "they are such dear little things." And then they are so plucky and determined in defending themselves, even against heavy odds, that one cannot help admiring this trait in their character.

But we must think of the task before us, which is to write something about Bantams that will be of use to our readers: and so we think that we cannot do better, in chapter 1, than consider what sort of houses and runs will do for them, and how they can be kept to best advantage, and with least trouble. Chapter 2.—On food and water, and modes of feeding. Chapter 3.—Proper size and weight of Bantams in relation to the larger breeds. Mating, crossing, interbreeding, and best time for hatching and rearing Bantams. Chapter 4.—Variety of Bantams, old breeds, intermediate ones, and newer varieties. Chapters 5 to 17.—Descriptions in detail of the different varieties, with standards for judging, etc. Chapter 18.—On preparation of Bantams for showing, sending to the shows, and treatment of birds on return from the shows. Chapter 19.—On diseases of Bantams, and treatment when ill. Chapter 20.—The Bantam Club, its formation, list of officers, etc., number of members, officers for the year 1892.

NEW VARIETIES OF BANTAMS
Produced by Mr. W. F. Entwisle

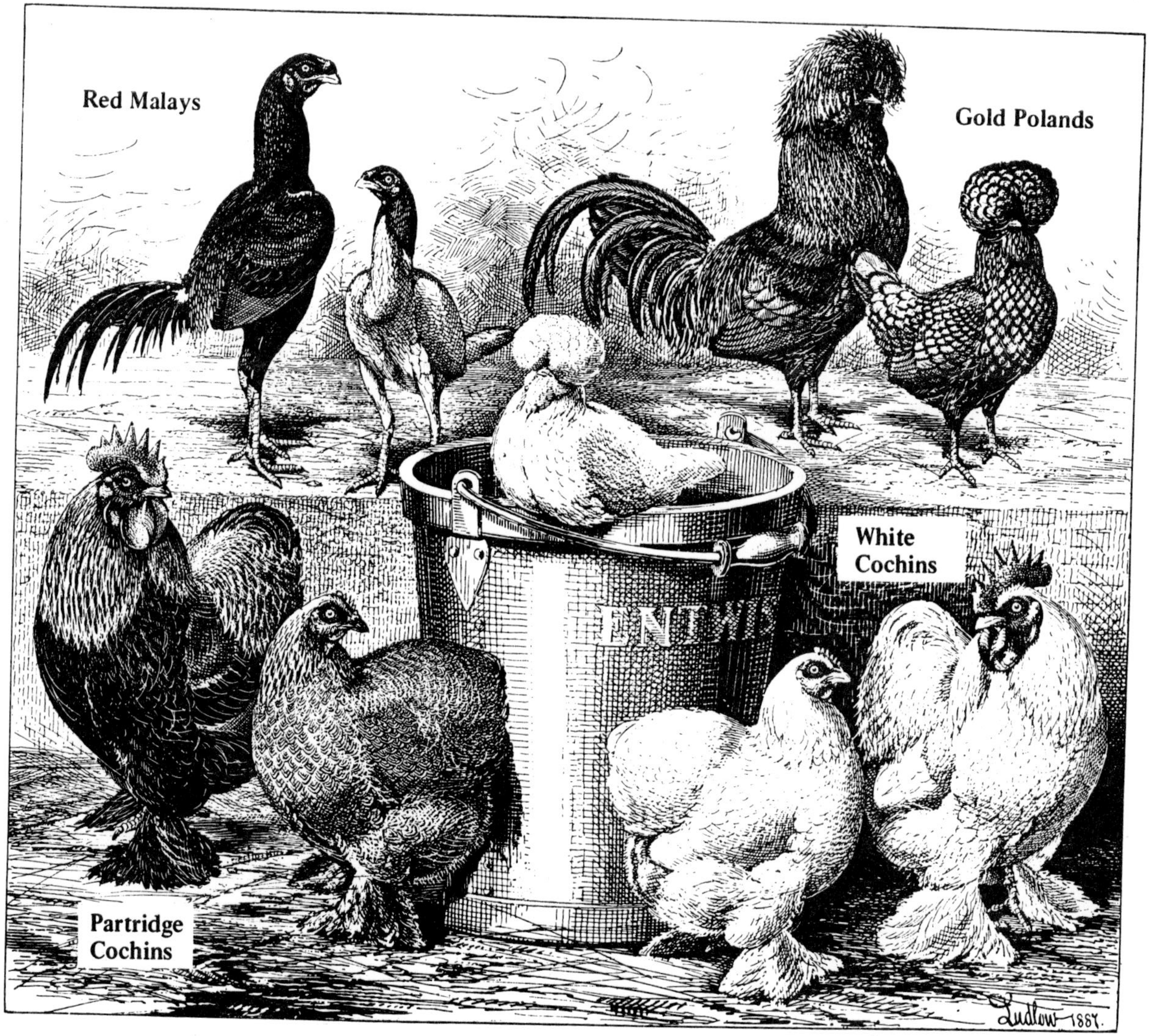

CHAPTER I.

HOUSES AND RUNS, ETC.

AS we remarked in the introduction, Bantams can thrive and do well in very limited space, and with very simple accommodation; and if any outbuilding, dry and free from draught, is available, there may be no necessity for providing a special Bantam house for one's pets.

There are, however, so many really well-made and admirably contrived houses now advertised for sale, at prices barely covering the cost of material and labour, that one can be supplied, at a day's notice, and at a cost of 30s. to 40s., with a really good and well fitted up house of almost any pattern, large enough for 12 to 15 birds.

We are using a number made for us by Mr. Wm. Calway, of Sharpness, Gloucestershire, and have been much pleased with them. They are neat and ornamental, easily moved by two men, well ventilated, fitted up with two perches, two nest boxes, a ladder, two doors, and underneath the floor a large space to be utilised as a dust bath and shelter for the fowls.

To these houses we have added wired-in runs of convenient size, so as to protect the birds from foxes, which abound here.

We have seen a very conveniently constructed house, built to stand in the centre of a square, providing four separate runs and houses. This is an economical plan, as it saves a good deal of woodwork.

There are also several very good lean-to houses offered at very small cost, easily fixed to any existing wall.

Whatever size and shape the house be, it is well for the floor to be raised so far above the ground as to afford dry shelter under it, and there to provide the fowls with their dust bath, which should be considered a necessity in every poultry yard, and be regularly replenished with ashes from the grate, finely sifted, and with the frequent addition of flowers of sulphur.

Where it can be done, it is well to have the run, or at least a portion of it, covered in, so as to keep it dry; and we know of no better contrivance for the purpose than the patent wire-wove roofing, which is durable and inexpensive, and gives light and ventilation. The feather-footed breeds, such as Cochins, Brahmas, Booteds, or Sultans, *must have dry runs* as well as dry houses, otherwise the foot feather can never be kept in good order.

Nearly all our Bantam houses are built of wood; some built together in pairs, with runs 12 feet by 20 feet, wired in, but made to open into still larger enclosed runs when we choose. Others are detached, which we prefer on the whole; though we will now describe one row of six houses and runs, each 6 feet square, and 7 feet high at the front, and $5\frac{1}{2}$ feet at back, which we have had in regular use for a dozen years at least, and found as convenient as any, especially for the feather-footed birds. This building, 72 feet long, was first entirely roofed with $\frac{3}{4}$ inch boards, over which was laid very heavy and thick felt, well tarred. The whole outside, woodwork and felt, has been again and again tarred at intervals of a few years, and the inside fresh limewashed every year. The front of the runs is boarded 3 feet high, and above that covered with 4 feet wide wire netting, $1\frac{1}{4}$ inch mesh. There is a small doorway from the house to the run, and a second doorway fitted with a sliding door opening upwards, leading into a large grass run. These sliding doors are fastened, opened or shut, by iron pins (attached to the doorposts by chains), fitting into small holes in the boards exactly on the top of the doors when shut down; and in fine dry weather the birds enjoy the run in the grass, but when rainy, or wet with frosts, they are confined to their covered-in runs and houses. The doors in this row of houses are all opposite each other, so that when all are open one can see right through them, or enter at one end and leave at the other end of the row. Each house has a window 2 feet square, the bottom of which is about 4 feet 6 inches from the ground; and a perch 6 feet long, fixed 18 inches above the ground.

There are in each house six nest boxes, 11 inches square, on the floor, quite at the back; the fronts are boarded up 4 inches high, the divisions are 14 inches high, so that the opening is 10 inches, and over the tops an 11-inch board keeps the whole row of nests clean. Our experience of these nests would lead us to adopt the same plan in future. Each little Bantam hen likes to have *her own nest* to lay in, and not to share it with others. As the whole row of nests can be distinctly seen from the window, we can better tell which hen lays in each nest, and if desirable, set her eggs separately, so as to know her own chickens. We now have all our nests made on the house floors, but all are made rat-proof by small mesh wire netting under the floors.

All these houses have earth floors, *i.e.*, they are built on the ground, and are without wooden floors, but they are covered several inches deep with fine sand, which is raked over every week, and passed through a fine sieve or riddle, and all excrement carefully removed to the manure pit. We have also quite a number of little houses or huts, each 4 feet square, 5 feet high at front, and 4 feet at back, with half the back opening as door, and a small doorway, 9 inches by 6 inches, in the front, opening into a wired-in run of 20 feet by 9 feet, and in some cases 20 feet by 6 feet, which we find very convenient for a breeding pen of a Bantam cock and two or three hens.

These little houses are fitted with sliding ventilators, on the same principle as those in railway carriages. We find this a very good and satisfactory arrangement.

Compact Poultry House.
(Mr. Wm. Calway, Sharpness).

The subject of perches must now occupy our attention, for they are of far greater importance than many people imagine. Every year many hundreds of good birds are completely spoiled through unsuitable perches.

Great differences of opinion are expressed as to perches, some liking flat, some round, others oval and rather broad. Some place their perches high up, others quite low, while others again choose a medium height.

The opinions we give are based on a lengthy and somewhat extensive experience of all kinds of perches, and places for them, and we think our hints on this head will be found useful.

For Game and Malay Bantams, where the formation of the foot, and especially the position of the hind toes, are matters of the utmost importance, it is essential that the perch shall be so small that their little feet can quite clasp it, thus preventing their becoming "duck-footed," which all Game fanciers know should disqualify any otherwise good bird. It is not, however, always easy to obtain perches strong enough, and at the same time slender enough, to secure this end. We have adopted two plans, sometimes one, sometimes the other, as follows:—Some of our houses were from 12 to 18 feet long where the chickens roosted, and we had large poles of those lengths resting on strong supports nearly 2 feet above ground, one at each end and one or two between, as we found requisite, and wedged so tightly that the poles would not turn round, or move at all. Then at distances of about 3 feet, we had supports screwed into the larch pole, supporting the slender perch, which was about 4 inches above the pole. Owing to these supports, the perches were quite firm and safe for the birds, and we never had any chickens duck-footed that slept on those perches. Our other plan was to suspend slender perches from the roof by strong wires, but we had several accidents through birds missing their footing in flying up to these perches, until we substituted wooden supports nailed to the roof, and made so rigid that they would not swing at all, then they answered quite as well as the others, and we found them less troublesome to fix up and remove. In all cases we prefer having perches for Bantams, from 18 inches to 26 or 28 inches above the ground, and then even the laying hens of the more delicate kinds are not troubled in flying up to them. For Rose-combed and Sebrights, and the other clean-legged varieties, small and slender perches are to be preferred, say 1 inch or 1¼ inches in diameter, and the birds are not likely to meet with accidents in flying up to a height of 24 or 26 inches above the floor.

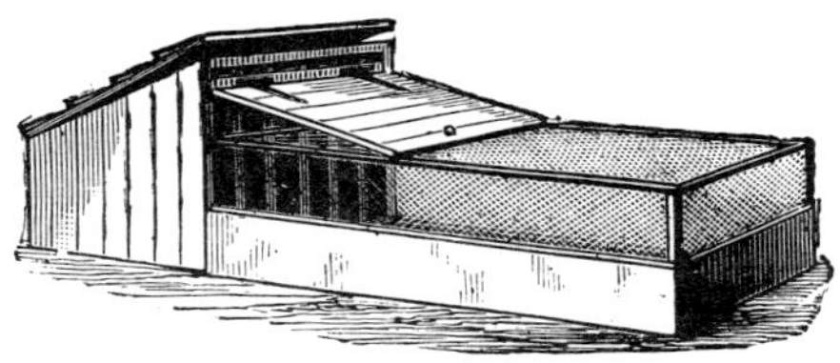

Chicken Coop, with Run.
(Spratt's Patent, Limited.)

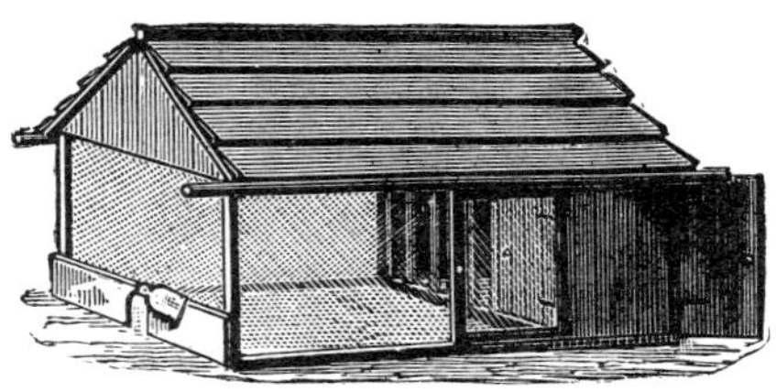

Chicken Coop, with Covered Run.
(Spratt's Patent, Limited.)

For the feather-footed breeds we prefer, and strongly advise, semi-circular or oval and rather broad perches, such as a larch pole 3 inches in diameter, sawn down the centre, and fixed flat side underneath, or 3-inch battens with the edges rounded off, and we have these about 18 inches above the floors of their houses.

Care must be taken to keep the perches free from parasitical insects; and we find frequent application of paraffin oil a certain preventive, and one which is easily applied.

For nests, no plan that we have seen adopted seems to answer better than to make them on the floor; and for Bantams, 10 or 11 inches square will be found ample room for the outside frame, and a piece of board 4 inches wide, the whole length of the nests, nailed on the bottom of the fronts to keep the nests in. The divisions may be from 11 to 14 inches high, ours are 14 inches, leaving 10 inches opening, but 12 inches will be found to give room enough. Then a board, or better still, two boards of 7 inches, so as to project a little over the top, will complete the nest boxes. We then like to fill them to a depth of 3 inches with earth, and shape them like a basin, filling the corners completely, and then give a lining of soft hay or straw.

As we usually have a nest-box for every hen, we like to let her sit where she has laid, unless she has troublesome companions, and then, of course, she is better removed to a quiet sitting-house. Our nests are always supplied with small wooden or earthenware nest eggs, as near the size of Bantam's eggs as we can find them; and we often put four or five in a nest to induce the laying hens to commence sitting.

We have often been asked what size of run we think necessary for keeping a pen of Bantams well and hearty through the breeding season.

The answer is not easy to give in few words. Much depends on situation and climate, and much on the variety kept, but still more on the amount of care and attention given to the birds.

We have often in this respect seen abundant proof of the old saying, "Where there's a will there's a way."

Not long ago we were asked by an exhibitor, whose name we had been familiar with for years, to look at his birds at home, and there, in a very small back-yard, perhaps 15 feet square, we saw six houses and runs containing six varieties of Bantams. These houses and runs were built two tiers high, or in upper and lower storeys. Every place was kept scrupulously clean, and though the houses were only a little over three feet square, and the runs not more than nine feet long, every bird looked bright and healthy. Though many of them were hatched and reared in those limited spaces, some of them had taken high honours in the show pen, including the Crystal Palace.

These birds were fed and cared for like ladies' pet cage birds; water cresses, or lettuce, or other green food was given daily in small quantities. Jenkinson's Perfect Gravel was given in little tin cans, and they had fresh food and water three times a day. Not far away from this fancier we called upon another in the same street. He kept all his Bantams in very small houses and runs, no larger than those just described. No grass runs, and all green food had to be bought for them or sent for from the country, yet we noticed every bird looked well, and some of them had so lived for three or four years, their lives occasionally varied by a journey to a show, where they generally took prizes. The main reason of success in these cases was the scrupulous cleanliness and careful feeding; and if a bird seemed to be at all out of sorts it had, at once, a "tonic" or a "reviver" given to it before it got seriously amiss.

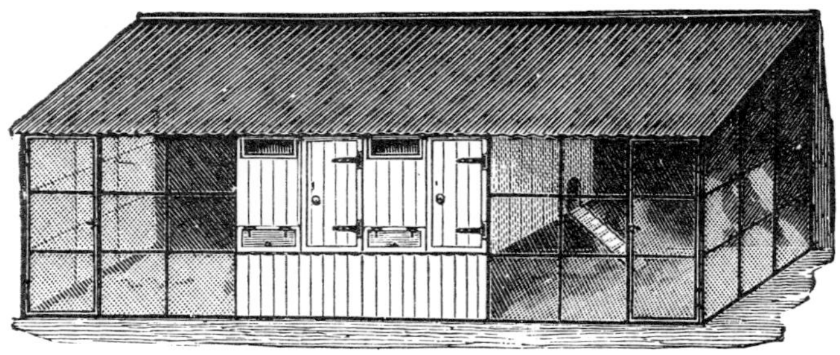

Lean-to Poultry Houses, with Covered Runs.
(Spratt's Patent, Limited.)

Still, though we could multiply instances of this sort, and in many parts of England, we prefer to recommend a more extended run, and ready access to grass where it can be given, thus saving time and labour.

If three or four Bantams are allowed the run of a kitchen garden, except where seeds have been newly sown, they will generally do more good than harm, and we may add that the small feather-footed kinds *scarcely ever* leave marks of scratching behind them.

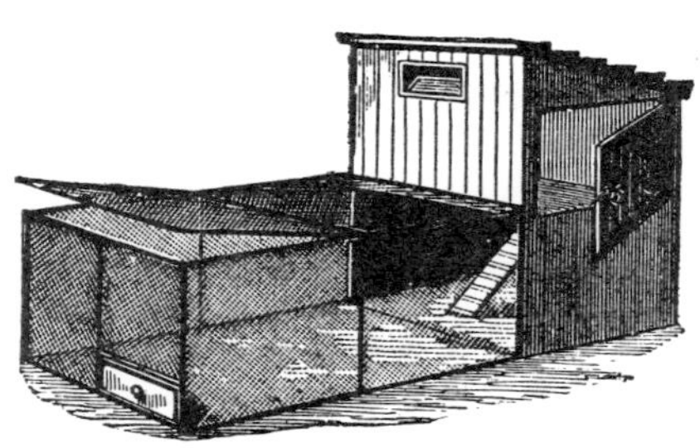

Poultry House, with Open Run.
(Spratt's Patent, Limited.)

Training cages or pens are needed by any Bantam keepers who intend exhibiting their pets.

These cages should be about 15 inches each way; and it is as well to accustom the birds pretty early to them, placing them in when hungry, and feeding them there, so that they will not be frightened.

The cages should be kept at a convenient height for observation, so that the birds shall be level with the eye; and they may be entirely of wire, such as Messrs. Spratt's show pens, or of wood with wire fronts, or only a wooden skeleton and bottom, wire front, and the sides, top, and back covered with canvas or calico.

On the special use of these cages we shall have more to say under the heading of Preparation for Showing.

Before we take our leave of this part of our subject (houses and runs), we must mention that there are many, yes, very many, true Bantam fanciers, who, for want of room at their own homes, have to keep many of their pets out on "walks," as they are called; sometimes a friend offers the necessary accommodation; more frequently the case is very similar to our own experience, as follows:—

When passing through the country, we have seen a detached cottage with a garden, and a few evidently well-cared-for fowls, and have ascertained that the people were trustworthy, and willing to keep only our Bantams, and perhaps a few hens for laying eggs, and when broody rearing Bantam chicks. We have then arranged for the dismissal of the other fowls, and installed a breeding pen of our own birds; and many a cottager has thus made double and treble the amount of the rent by rearing broods of our Bantams. Several of our largest and most successful breeders are glad thus to avail themselves of country "walks."

In these cases, there is generally some dry and sheltered outbuilding available for the roosting-house, and for the hens' nests, but sometimes we have sent our own little movable house and run combined, each 4 feet square, and the run wired over the top, so that the birds could be safely shut in at night and let out in the morning when the family were about; and some of these little houses have been in regular use for nearly twenty years.

Coops for the chickens are of almost as much importance as the houses for them when upgrown, and there are many most excellent designs of coops constantly advertised in the poultry papers, so that one need have no difficulty in selecting size and kind suited to one's requirements and locality.

Cats and rats are often the most dreaded enemies of the young chickens, but there are plenty of coops proof against both cats and rats. We will describe one such coop of a pattern we have a few of, and find very useful. Length 5 feet, depth front to back 2 feet 6 inches, height at front 2 feet 6 inches, height at back 1 foot 9 inches, boarded roof, back and ends; and the roof also covered with a stout waterproof cloth. The whole painted dark green, the floor covered with fine wire netting ¾-inch mesh, the front boarded 1 foot high, then ¾-inch wire netting to the top. One end of the coop (2 feet) has boarded floor, and is closed in by a door which fastens with a button. At the other end of the coop is a small sliding door which closes a doorway about 2 inches by 2½ inches, quite large enough for chicks to run in and out, but too small for the mother to pass through. This small door can be opened by day, and closed any time when desired. We hardly ever lose chickens from these coops, and intend to have more of them.

Shelter is needed by all Bantams, whether young or old, against sun, wind, and rain or snow. Against the too powerful rays of the sun in hot weather, various kinds of screens have been provided; these we can only hint at in passing, but very open and coarse canvas is often as useful as any material. Trees and shrubs planted in their grass runs are of immense benefit to the birds, and we have often found great benefit from ordinary hurdles covered with leaves or straw, reared up in such a manner that the chicks could get under the shade.

Hurdles so covered, and placed a little height above chicken coops when out in the open, so as to admit a free passage of air above the coop, are of much use both to the hen and her chickens.

Shade of some sort should be provided for all Bantams to prevent injury to their colour during the hot sunny weather, for not only are Whites tanned, but all colours are more or less injured, even black turning brown.

And in Winter, and the early Spring, it is equally important that shelter and protection be given from the biting east winds, and from damp in any form, or the birds are sure to take cold, and probably have attacks of serious illness.

As prevention is better than cure, don't neglect the above precautions.

CHAPTER II.

FOOD AND WATER, AND FEEDING, ETC.

BANTAMS should be *well fed*, and with *the best of foods;* by this we do not, of course, mean overfed, and this we believe is the mistake most commonly made. The food should never be given in such quantities that the birds' appetites are quite satisfied before all the food is eaten, and so a quantity is left on the ground. This is waste, besides being injurious to the birds.

They should not be fed too frequently; twice, or at most three times a day is quite often enough for the adult birds; but chickens require food a little at a time, and often. We think no better plan can be given, as a general rule, than to give a feed of warm meal as early in the morning as possible, a light feed towards noon, and a more liberal feed of grain about an hour before the roosting time in the evening; these hours must of necessity vary according to the season of the year.

If the Bantams are in confined runs, more must, of course, be provided for them, and they will then have to be supplied daily with green food of some kind; grass and clover we believe are the best, then dandelion, lettuce, cabbage, cress, etc., in the order named. Apples they are very fond of; and pears, where they are plentiful and can be spared, are good for the birds. Such roots, too, as mangold wurzel, swedes, and beetroot, are very good for them.

We prefer giving all such foods and any table scraps at noon, or near mid-day. And for hard grain, wheat, short heavy oats, barley, rice, or dari, small maize, and occasionally buckwheat, though many Bantams will scarcely touch it even when hungry. As to quantity, it is safe to give an ordinary handful to every three or four Bantams, and it is well to vary the grain occasionally, though fowls in health never tire of wheat, and it is the grain we use most of. Good heavy short oats we take to be the second best; but always avoid long thin light oats as worthless, and, in fact, injurious. Bantams always like rice and dari; sound rice good enough for fowls can generally be bought at about 11s. per cwt., and we think then it is a valuable food *in its place*, but it cannot take the place of such grain as wheat or oats. The following table will show the relative value per cent. of some of the best foods:—

Grain and Meals.		Flesh forming material. Per cent.	Bone forming material. Per cent.	Fattening and warming material. Per cent.
Wheat	contains	12	2	73
Barley	,,	12	2	61
Good Oats	,,	15	2	53
Buckwheat	,,	12	2	64
Maize	,,	11	1	73
Rice	,,	7	less than 1	80
Dari	,,	8	1	76
Oatmeal	,,	18	2	69
Fine Sharps	,,	18	5	59
Barley Meal	,,	11	2	62

Thus it will be seen that such food as rice, dari, and maize have only half, or less than half, the bone-forming material of the other grain, and need to be counterbalanced by such meal, in their soft food, as fine sharps, which contains an excess of bone and shell forming material.

Where fine sharps, of really good quality, can be obtained, it is one of the best meals that can be used, especially to mix with oatmeal or "Liverine." This Liverine we must say a few words about, as we find it the best substitute for insect food we have ever tried.

It is prepared by the Liverine Co., Limited, of Grimsby, from fresh cod livers and other fresh and sound fish, cooked, kiln dried, and ground into a powder, with oatmeal and other good sound meal; and it is sold at about the same price as oatmeal. We have used it from the time it was first brought out publicly, and we give it to our breeding stock, and to the young chickens also. It is very rich in nitrogenous substances, and in phosphates; it contains the fish *bones* thoroughly pulverised, and the fowls all relish it amazingly; so do pigeons, cats, dogs and other animals.

We mix one-third Liverine, one-third fine sharps, one-third oatmeal, with boiling water into a stiff paste, and then sprinkle a little dry oatmeal into it till all becomes quite crumbly, and then feed with it while warm. Condiments and spiced foods are right enough in their place, and especially during moult and in cold damp weather, or during east winds; but we think at other times the more sparingly used the better.

There are many makers of them who use pretty much the same ingredients, and they are so well advertised in the poultry papers that we need not further refer to them here. It will be seen that Liverine is a totally different article, and we think far preferable to Prairie Meat Crissel.

In choosing grain for Bantams, always buy the best; for so-called *cheap* grain, or cheap mixtures for "fowls' corn" are often very dear in the end.

GRIT.

In order that the corn and other food may be thoroughly digested, sharp grit is absolutely necessary, and should be constantly at hand. We supply each of our houses with a tin or jar full of it, and keep replenishing every week. Lime in some form is also necessary. Old mortar does very well.

It should not be necessary to say that in hot weather the Bantams need less food than in cold—it is so, we know, with ourselves. In cold weather we need so much food, and more animal food, to keep up the heat of the body, whilst in summer this is supplied by the heat of the sun.

In very cold weather, therefore, give the Bantams a little extra food, and of the more generous kinds.

WATER.

Fresh, and pure water, though we mention this last on the list, is of the greatest importance, in order that the birds may be kept in health. Where the fowls are not confined, but have access at all times to running brooks or springs, no trouble is necessary on this point, but as such cases are the exception, and not the rule, we feel that we must try to insist upon the greatest care being taken about the drinking water, and water vessels.

In hot weather never allow the water vessels to stand in the sunshine—always in the shade. And in winter, when frosty, they should only be half filled each forenoon, and emptied before dark in the evenings, to prevent bursting by the frost. In most Bantam houses, drinking fountains are provided—some of metal, but probably more of earthenware; and these, though more liable to be broken, are, we think, to be preferred, especially when made in two pieces, and in such form that they can easily be scoured and cleaned inside. It is well occasionally, and especially in the early spring and autumn, to put a small piece of sulphate of iron, and a few drops of sulphuric acid—three or four drops of acid, and a piece of the sulphate of iron the size of a hazel nut, to a quart of water for the birds to drink. This must not be continued too many days together, but in bad weather a week will be sufficient at a time.

Before closing this chapter, we must give some few directions as to the special feeding of Bantam chicks, as we don't think it wise to feed all breeds alike. If we study the points we most aim at, it should not be very difficult to understand that in such breeds as Game and Malays, where length of bone is required, and *strong bone* (we mean bone of good quality, not great coarse thick bones like those of the cart-horse, but the finer quality of the race-horse), we must feed with more bone forming meals—such as the best fine sharps—than we would give to such breeds as Pekin, or any of the other shorter legged breeds. And again, the harder feathered Game Bantams should not be forced to mature fast by too stimulating food; if they are, even the best-bred birds will only produce chicks like the monthly roses—very beautiful just before they are in full bloom, but soon done for.

Care must, therefore, be taken to give suitable food for producing the points desired; and as our own system has been successful we will explain it.

When the chickens hatch, we leave them in the nests for the first eighteen to twenty-four hours, only removing the egg shells; then the chickens should be strong enough to be taken from the nest and fed. First give finely chopped, hard-boiled yolk of egg (not the white, for that would be useless for the chicks—only the yolk), mixed with about double the quantity of bread crumbs. That may be given six times for the first day's feed, and the same the second day; the third day add equal bulk of rice, cooked with just as much milk as it will take up and still remain in single grains; and broken rice of good quality is better than whole.

Feed every two hours. From the fourth day to the eighth, we feed chiefly upon the cooked rice, and discontinue using egg food after the third day; instead of it we give Liverine, sharps and oatmeal. This food will do for all kinds of Bantams alike for the first five weeks. It is safe at a fortnight old to give a little wheat, buckwheat, or canary seed, always remembering sharp grit.

When the chickens have got their first set of feathers on their backs, breasts and wings, say from five weeks old, the proper course of feeding must be carefully carried out, giving more rice, dari and oatmeal, and occasionally buckwheat to the breeds we want to keep short-legged and small, but plump and feathery; and more wheat, sharps, oatmeal and Liverine to the harder feathered, and longer legged breeds. One more hint, and then we must leave this subject. Where practicable, give a wide range and plenty of exercise to all Game Bantam chickens; make them *seek* their food, and *work for it*. Corn scattered wide among the grass will do this as well as anything will.

The shorter legged and softer feathered breeds need less exercise; but still they must not be without some, or they will not keep healthy.

CHAPTER III.

THE RELATIVE SIZE AND WEIGHT OF BANTAMS; MATING, CROSSING, AND INTERBREEDING, AND THE BEST SEASON FOR REARING THE CHICKENS.

WE are constantly meeting people whose ideas of Bantams seem to be pretty correct generally, but on one point we differ seriously, viz., the correct size and weight. They appear to think that all Bantams ought to be of one size and weight, or nearly so, whatever the difference be in size of the large breeds of which the Bantams are miniatures.

Now, this does not appear to be reasonable; nor do we consider it desirable that all should be brought down to one level. In point of fact, the thing will prove to be impracticable. We have taken the opinion of several of the most competent of our judges, and also of several well-known successful breeders and exhibitors, and the result is, that by a large majority, we have agreed that one-fifth the standard weight of the large breed should be considered the proper standard weight of its miniature. Thus the Game cock 7 to $7\frac{1}{2}$ lbs. would give us 1 lb. 6 oz. to $1\frac{1}{2}$ lbs. for the Game Bantam cock, and the Bantam hen 4 to 6 oz. less. The Malay cock $8\frac{1}{2}$ to 10 lbs. weight, the Malay Bantam cock $1\frac{3}{4}$ or 2 lbs., Malay hen $6\frac{1}{2}$ to $7\frac{1}{2}$ lbs., Bantam about $1\frac{1}{4}$ to $1\frac{1}{2}$ lbs. Black Hamburgh cock about 5 lbs. or a little over, Black Rosecomb Bantam 1 lb. or an ounce or two over. We merely give these few illustrations, but if others are noted, it will be found that practically the one-fifth weight corresponds with the most correct judging of the present time. Weight alone, however, would in many cases be most deceptive, for a bird may be unduly fat, and so overweight, or unusually thin and light, and consequently under the different conditions the same Bantam might vary as much as six ounces. Our remarks about weight must, therefore, be taken to mean weight of the bird when in good firm flesh, and show condition.

There has, of late years, been a tendency among breeders to try *how small* they could produce their Bantams. We can follow them a long way in this direction with approval and pleasure, but we have already gone too far with one or two breeds, and unless a reaction soon sets in, those breeds stand a good chance of being utterly ruined. We really think eight ounces far too small a weight for any Bantam hen, if she is to be of any use; but that is about the actual weight of a very noted winner, which, according to our one-fifth scale, should score about fourteen ounces.

Mating must now occupy our attention for a little space.

The number of hens mated to one cock will vary much, according to the variety and other circumstances. Two or three hens will be found the best number for such breeds as Sebrights, Japanese, Burmese, Pekins (if imported), though the English strains are hardier, and an extra hen may be added. But the hardier breeds are generally mated in sets of five or six hens to one cock. We think it best to keep the birds mated together all the year round, and from year to year, too, when they are found to breed well together. Most Bantams are pugnacious, and resent the intrusion of any fresh arrival, so that the new comer won't have a very pleasant time of it for a week or two, and laying will be considerably retarded.

Where early chickens are wanted, the birds to be mated together should become used to each other in October or November at latest, so as to have eggs in January. Many people do not care to "put up their pens for breeding" till January, or even February. They don't expect very early chickens.

Too much care cannot be exercised in mating

the birds, that the hens and the cock are suitable for mating together; not too closely related to each other in blood, nor yet totally unrelated, in which case one has to wait and see how things will turn out. In the case of breeding for feather points, many an inexperienced breeder has bought a cock of one strain and mated it with hens of quite another strain, on both sides perfect specimens of the same breed, and the result has been a most disappointing failure, and the dispersal of all the lot, old and young; the reason being that the cross of totally unrelated strains, each carefully bred, may be, but upon different lines, always produces more cases of reversion, and some of them very curious ones, than the mating of birds somewhat nearly related to each other. Now, in the supposed case under consideration, had the old cock been mated the second year with his own pullets, and chickens been reared from them, and the old hens been mated with a cockerel from one of these pullets, or *vice versa*, there would be much more uniformity in the chickens produced. Interbreeding, to a great extent, is the means found most reliable in fixing a stamp of uniformity upon a strain.

The extent to which in-breeding may safely be carried, varies very considerably in different breeds of fowls, and quite as much in different strains of a breed.

In the most carefully bred strains, and where the best sanitary precautions have been adopted, and no hereditary weakness has been manifested, in-breeding may be carried on for many years without apparent injury—our own experience proves this most fully.

But there are times when crossing becomes necessary, or at least desirable; then, if you know that the bird you cross with is of a pedigree that will stand the test of careful investigation, even if the first cross should look unsatisfactory, persevere, and breed back on both sides, son to mother, or father to daughter, and then continue in-breeding, and the result will come right in the end. This is a most interesting subject, and we will give a few practical illustrations under the heads of some special breeds, further on.

We now come to consider the best time for rearing Bantam chicks.

Of course, this will depend mainly upon the purpose for which they are wanted. If it be merely to keep them as pets at home, or ornaments in the grounds, then we should say set the eggs about the middle or toward the end of March, and the chickens will be reared at a time when the weather is all in their favour, and they will give the least trouble.

If the question be asked, what time of the year may we expect that the weather will probably most conduce to develop the *desired shape and style*, we should have to answer, the early spring, for all long legged varieties, and August or September for hatching the shorter legged breeds, as the cold weather coming on when their limbs should be growing, they crouch down under the hens, or huddle up together, in order to keep warm, and so become more dwarfed in stature and style, while the early hatched birds continue to grow more and more limb as the weather gets warmer, and insect life (bone-forming food) becomes more plentiful.

If it be a question of rearing Bantam chickens for prize competition at shows, then, of course, the earlier they are hatched the sooner they are ready for the summer shows; but for the "big events," the Crystal Palace and Birmingham Shows, the chickens need not be hatched before May or June. Probably April and May will be found the most convenient time for hatching most of the chickens.

With reference to Game Bantam cockerels, which are usually dubbed when shown, they should be six months old at the least before they are dubbed, and it is better to keep them at home for ten days or a fortnight after that before sending to shows. Game Bantam pullets are never smarter looking than between the ages of five and seven months.

Some other varieties improve in appearance as they mature and "thicken out," as some express it; the Sebrights, for instance, improve much in carriage, so do the Pekins and Brahmas.

But we must pass on to our next division of the subject.

CHAPTER IV.

THE VARIETIES OF BANTAMS.

SOME of the varieties of Bantams are known to us as *old* breeds, such as the Rosecombed Blacks and Whites, Sebrights, Pheasants, or Spangled, Nankins, Booted, etc.; others as *intermediate*, such as Game, Cuckoos or Scotch Greys, Frizzled, Rumpless, Japanese, Buff Pekins, and *newer* importations or productions, such as Burmese, Black Pekins, Cuckoo Pekins or Cochins; also, Partridge, and White ditto, Malays, Indian Game, Aseels, Brahmas, Polish, Sultans, etc., etc.

In thus attempting to classify the varieties under three heads, we feel that many fanciers are likely to think, if not to say, that two divisions would be quite sufficient for the purpose. Perhaps so, but we think no harm can be done by attempting to define a little more distinctly which have been the longest known in this country; which have been produced or brought into notice by the oldest fanciers of the present generation in their younger days; and, lastly, those which have been introduced to the public within the last ten to fifteen years as new breeds.

Without doubt, everyone of those we place as the oldest known breeds (excepting Nankins and the old-fashioned Pheasant Bantams) has greatly improved in points during the last forty or fifty years; so much so, that if we had one of the best specimens of the year 1840 placed alongside one of the winners of the 1890-91 season, the difference in value would be greater than that between 1s. and £1—at least, if we except Sebrights, that would be the case; but in the laced Bantams some good points have deteriorated, whilst others have considerably improved.

We think that we shall be quite within the mark in saying that not one of those we have placed as Intermediate was known in England fifty years ago, certainly not sixty years ago; and there would not be very much difficulty in defining the localities where each first made its appearance. For instance, there can be little doubt that to Mr. John Crosland, of Wakefield, belongs the credit of the production of the earliest Game Bantams. To Mr. Tegetmeir the earliest production, and for many years the careful cultivation, of the Rumpless Bantams, which are to be seen in so many colours, the latest variety of all being perfectly marked silver-laced, and perfectly rumpless, bred by Mr. Garnett, and exhibited for the first time at the Poultry Club's Show in London, 1889. This, we may add, is not a solitary specimen, Mr. Garnett having bred others.

We have failed hitherto in tracing the Cuckoo or Scotch Grey Bantams further back than about thirty years, when the first we saw were bred in Scotland, and were as well marked as if the breed was an old-established one. We think, however, that if it had been longer known in Scotland, we should have heard something of it, for we made very diligent inquiries in various directions down in the North.

Buff Pekins were first introduced into England about thirty years ago, and were in-bred from necessity for many years until they had become nearly extinct through sterility.

Japanese were also introduced to this country about the same time, but in small numbers, and nearly all of one colour—nearly white. Later importations have, however, furnished great variety in colour.

Among the newer introductions from abroad, the chief are the Black Pekins; they are now firmly established in this country and are great favourites; and next, we would place the Burmese, which hitherto have proved too delicate to do well in our damp and variable winter climate. The others mentioned are home bred, the best known being the Malays, which are making their way, and becoming favourites with a goodly number of fanciers.

But as we propose to take all these varieties in turn, and go more fully into their history and points, we must not take up more time in this chapter, but at once proceed with the descriptions, commencing with Game Bantams, and then following with Malays, Aseels and Indian Pheasant Bantams (the miniature Indian game fowls), then Cochin Bantams, etc., etc.

CHAPTER V.

GAME BANTAMS.

THESE are undoubtedly the most popular, and the most extensively kept of all the Bantam tribes. The best known colours are the Black-breasted Reds, commonly known as Black-reds; Brown-breasted Reds (abbreviated to Brown-reds), Piles, Duckwings, Wheatens, Whites, and Birchen Greys. There are also (or there were) Blacks and Blue-duns, but we have not seen any of these two latter for some years, and fear they have become extinct; they were never popular as show birds.

Of the above colours the Black-breasted Reds are the chief favourites, and present the least difficulties in breeding true to colour; next as favourites come the Red Piles, then the Duckwings, and then the Brown-breasted Reds, which are still in few hands, though we hope there will be a change in this respect, two or three spirited fanciers having commenced in good earnest to try and improve the breed, and, if possible, wrest from Messrs. Dan Clayton & Stretch the position they now hold as chief winners in Brown-reds.

In all colours of Game Bantams the chief points are style, good colour, hard feather (*i.e.*, narrow, short, small, and closely-fitting feather throughout), and small size.

With most judges and breeders style is considered of the greatest importance, and so we place this first. By style we mean, or include, the carriage and the general shape of the bird, which should be erect, tall, smart, and racy; the head long and narrow, forming with the neck a graceful curve; the neck itself long and rather slender (not "throaty" nor "bull-necked," and full and thick in hackle, which is a very great fault); the neck feathers cannot be too narrow and close-fitting, thus greatly adding to the apparent length and slenderness of neck, and showing off the shoulders to greater advantage. These should be carried rather prominently forward, and be broad and square; the back flat and short, and tapering off to a very fine, narrow stern; the chest rather broad, but not too prominent; the thighs long, muscular, and set well apart, and rather forward; the shanks long, straight, and slender, and as round as possible (not flat on the front nor square, both of which are signs of weakness); the scales on the legs small, fine, and close-fitting; the feet and toes sound, the toes being long, straight, and well spread out flat on the ground; the hind toes a good length, set on low, and exactly opposite the middle toe, so as to give a firm foothold. The hind toe is of the greatest importance, and should always be carefully considered in the breeding pen, as any fault here is almost sure to be transmitted to the progeny. If set on too high, the toe is usually too short, and only the tip of it touches the ground; that would be considered a bad foot, but not a disqualification. Very often the socket of the hind toe is faulty, and the toe itself grows sideways, and bends forward and inwards; the bird is then said to be "duck-footed," and that is a disqualification. The breast should be firm and meaty; the sides of the body well rounded, and the wings short and well curved, fitting closely to the sides of the body; if carried too high, and meeting across the back, the bird is said to be "goose-winged," and both this and the opposite extreme, wings carried low, and drooping, are great faults, which must not be tolerated. Many Game Bantams are too long in the wings, and also "flat-sided" (*i.e.*, the wings long and straight, without the well-rounded curve, which enables the wings to clip the body tightly); flat sides are to be avoided in all Game, both large and small. The tail must be small, and carried tightly together, and at a very slight elevation; the tail proper (*i.e.*, corresponding to the hen's tail) should not have more than 14 principal feathers, and they should be narrow and short; and the cock's sickles should be as fine and narrow as possible, a small "whip

BLACK-BREASTED RED GAME BANTAMS.

Bred by, and the Property of, the Rev. F. Cooper, Lea Marston, Birmingham.

Cock, winner of First and Cup at Crystal Palace, 1893. Hen, Second Prize at Crystal Palace, 1892.

tail" adding greatly to the appearance of a good bird. Besides this, we require in all Game Bantams good, large, bold, prominent eyes; in Brown-reds, as nearly black eyes as possible (they are of various shades of brown, but the darker the better); and the same applies to the Birchen-greys; all the other colours should have bright red eyes—some say ruby red—and on the whole we prefer this term, as it is understood to mean a clear, distinct bright red; others say blood-red is the proper term. We also look for fine skin, as smooth as possible on the face and throat, with thin, small, neat, well serrated, perfectly erect, and straight combs; small ear lobes, quite free from any white, though this is a difficult point to secure, and the general practice seems to be the removal by scissors of any white from the centre of the ear lobes.

It is, however, far better to breed out the faulty white lobes; *it has been done* in some of the best strains, and with care, white lobes should be soon a thing of the past in all Game Bantams. Hardness of feather is desirable in all the colours, but is more rare in Duckwings and Birchens than in any other colour. Hardness is partly the result of feeding, but more of careful breeding to secure this end. It is especially necessary that the cock bird should excel in this point; but this hardness is usually accompanied with a deficiency in colour, the hardest feathered, as a rule, being far too dark; so it is often necessary to mate these cocks with hens lighter in colour than the standard for shows requires, in order to produce bright coloured chickens.

Size is the next point to be considered, and our remarks on a former page will show that we are not in favour of excessively small birds; the desire to produce the *smallest* has rendered some strains almost worthless; they have become so delicate that they are only fit to be kept as cage pets. The best test for size is the eye, and *if the birds look small enough*, then the more they weigh and the harder they feel, the better, provided they are under 18 oz. as pullets six to nine months, and 20 oz. as hens; 20 oz. as cockerels of six to nine months, and 24 oz. as cocks; Game Bantams about two ounces off these weights, often appear really small birds. Another couple of ounces can easily be taken off such birds in training them for show, but the weights given above are those of birds in high condition for breeding.

We have now to describe colour; and as this point alone counts 15 to 20 per cent., it must be carefully considered. The Black-red cock should have face, throat and top of head a good healthy looking bright red; if not dubbed, his comb, wattles, and ear lobes should all be of the same colour (bright red). Beak, dark horn colour; his eyes, clear, full, rich red, or ruby-red—dark brown eyes, or yellow, or pearly (daw eyes) are very objectionable both in cocks and hens, and practically often disqualify. His neck hackles should be clear orange, or light orange-red (the fashionable colour now being considerably lighter and brighter than it was twenty or thirty years ago), the back and wing bow, bright crimson (or, in many birds, rich orange red, but crimson is preferable), the saddle hackles orange, to match the neck. The wings should have the butts black, shoulder coverts and wing bow crimson, wing bars steel blue, flight coverts clear rich bay or chestnut, breast and thighs bluish black, belly black, and free from any rusty colour, tail glossy blue-black, legs and feet willow or olive.

This is the proper colour for the show pen, but it will often be found that a cock of this colour, though breeding cockerels like himself, breeds nearly all his pullets foxey or rusty on the wing, and rather pale in breast.

Generally the best pullet breeding Black-red cocks are much of one shade of red all through, from head and hackle down the back and the wing bow, and shading off to slightly more orange colour on saddle hackles. This description of cock is not a favourite in the show pen.

The standard coloured Black-red hen must match the cock in colour of eyes, beak, legs, and feet, and have red face, comb, wattles, and ear lobes; the feathers on head and hackle golden, with narrow but well-defined black stripes on each side of the shaft, and rich golden edge all round the black; her back and wings, and the outer feathers of tail, should be all of one even shade of colour, a light brownish drab, each feather very finely pencilled all over and perfectly even with black; the effect being that at a short distance the colour appears to be plain brown. If the pencilling is too large in pattern it is termed "coarse," and considered very objectionable. Any red, foxey, or rusty coloured feathers on the wing denote a cross with the Wheaten-reds, and though birds of this description are among the best of cock breeders, they fail in colour as exhibition hens. The tail of the hen should be black, but if the upper and outer feathers show the same colour as the body and wing, it is an improvement rather than otherwise, and denotes purity of the pullet breeding strain. The throat should be light salmon, the breast rich salmon red, with the shaft of the feather a shade paler, and shading off to ashy-grey on the thighs and belly. Hens of this colour (which is the standard colour for showing) can breed perfect coloured cocks if mated to well-bred cocks of equally sound colour. Many breeders, however, for breeding their cockerels, favour the lighter shades in hens, with more or less of Wheaten blood. These are paler in breast, less striped in hackle, lighter in back, and with foxey or wheaten colour on the wings, and generally they show more style and harder feather than the best coloured hens do.

The Brown-red cock should have a dark gipsy face (*i.e.*, dark mulberry or purple), the darker the face and eyes the better; if not dubbed, the comb should be of the same colour as face; wattles, dark red; beak, the blackest horn colour; legs, feet, and toe nails, the darkest bronze, nearly black; head and neck feathers, lemon, striped with black, each feather having broad margin and shaft of lemon or straw colour, with narrow black stripes between; back, wing bow, and saddle, lemon, rich and bright in colour; breast, rich black, each feather showing pale lemon or straw-coloured shaft, and margin all round, giving a beautiful laced appearance; the lacing may come down on to the thighs, and all the rest of the feather should be black, and as bright and lustrous as possible. The above has been the fashionable colour for several years, but it is seldom we find in this colour the hardness of feather which is desirable, and, consequently, there are good breeders and judges who would substitute golden for lemon, or even go as far as to advocate bright orange; but, whichever shade it be, the laced breast and hackle must be clear and distinct. There are, again, those who like a red face or purplish red face in Brown-reds, but a very large majority prefer the darkest gipsy faces, which are much easier to obtain in hens than in cocks. The Brown-red hen must have a gipsy face and comb, and the darkest shade of brown eyes, nearly black; her legs and feet must also match the cock's; and the beak be black, or the blackest horn. Her head and neck hackle straw colour, with a narrow black stripe on each side of the straw-coloured shaft of feather. Pale golden hackles are next to straw colour in value, but dark or coppery hackles (which passed muster twenty years ago) are now quite "old-fashioned" and disliked; the breast must be beautifully laced with the same shade as the hackles, and all the rest of the feather must be sound green-black if intended for the show pen.

Birchen-greys are so very evidently first cousins to the Brown-reds, that we will take them next; and very few words will suffice to describe them, as they are exactly the same in markings as the Brown-reds, the only difference in colour being that the Birchen-greys are black and white where the Brown-reds are black and straw, or lemon.

Duckwings are most beautiful in colour when up to the standard requirements—viz., the cock's head and hackle feathers, creamy white, and if perfectly free from black stripes all the better. Back and wing bow, clear and even orange, or golden yellow, shading into straw colour on the saddle hackles; wing butts, breast and thighs, sound blue-black; belly, black, free from grey or coloured feathers; wing bar, steel blue; flight coverts, clear white, free from red or sandy colour, but each feather having on the end a blue-black spot, the whole wing when closed looking as if there was a bar or stripe of blue-black above the white, and the more clearly defined this is the better. Nearly all Duckwings fail in this point. The tail should be blue-black, and sickle feathers quite free from any other colour, either in their shafts or edges; the face and eyes as bright red as in the Black-reds and Piles; legs and feet, willow or olive green. The Duckwing hen must match the cock in colour of face, eyes, legs and feet; her breast must be salmon, like the Black-red hens, shading off to ashy-grey on the thighs and belly. Like the Black-red, she must be very finely and evenly pencilled with black, but on a French-grey or very pale slate-coloured ground; her head and neck hackle white, striped with black, like the Black-reds; tail, black, the top outer feathers marked like the back and wings.

The Silver Duckwings are just like the Golden Duckwings, except that the Silver cock has white hackles, back, saddle hackles and wing bow, and the Silver Duckwing hen is of a lighter shade all over than the Golden Duckwing; they are a handsome variety of Game, and breed very true, if not crossed; but they are often crossed, both with Black-reds and Golden Duckwings, to improve colour in the latter breed.

Red Piles, commonly called Piles, are very handsome. They have red faces and eyes, and either yellow or willow legs and feet—yellow preferred; the beak must correspond with the colour of the legs. The cock should be the counterpart of the Black-red; only, instead of being red and black, he is red and white, and as any red feathers in the breast of the Black-red cock are considered a blemish, so they should be in a Pile. He should be a pure, milky white in breast, thighs, belly and tail, white on the wing butts, and white on the wing-bar; his head and hackle orange red, as free from white as possible (though there is generally a little white in the lower part of the hackle); back and wing-bow, rich crimson, shading into orange or orange-red on the saddle hackles; flight coverts, rich chestnut or bay. If the tail feathers are slightly "peppered" with black, it is not considered much of a fault, if any; such birds have generally the hardest and best quality of feather.

The Pile hen should have red face and ear lobes, and red eyes; golden hackle—*i.e.*, as golden as possible, each feather has a white stripe down its centre, and in many hens the hackle is more white than gold, but it should be otherwise; her breast rich salmon red, with a lighter salmon stripe down the shaft of each feather, shading off to creamy white on the thighs and belly; her back, tail and wings should be a clear creamy white—as white as possible; but many very stylish birds are "rose winged"—*i.e.*, have rusty red markings on the wings, like the cock breeding Black-red hens have, and these hens often win, through superior style, in spite of the *faulty* marking, for faulty we must call it, however much some people may admire it.

A great number of Lemon Piles are bred every season from the very best coloured birds, and though these are sometimes useful as stock birds, they are not welcomed in the show pen.

The Lemon Pile cock is pale lemon in hackle, orange or lemon on back and wing-bow, pale lemon in saddle hackles, and the rest of his feathers white, or sometimes slightly coloured with chestnut on the flight coverts.

The Lemon Pile hen is a more pretty bird than her male companion. She has pale lemon and white hackle, pale salmon breast and all the rest milk white, and generally very good orange or willow legs. Such hens are of great value in breeding.

White Game Bantams should be very pure white in feather, with good yellow legs, feet and beaks, red faces, and red eyes. The Whites originally sprang from a strain of Brown-red Bantams, which we commenced to make about the years 1862-63, and into which we bred a cross of apparently pure Black Game, afterwards finding that this Black Game was herself the result of a cross with White Game, and it is a fact, well known to all breeders of Brown-red Bantams, that to this day pure white chickens come out of nearly every large hatch of Brown-reds. Black Game Bantams we produced in the same way, and by the same cross. In colour these Blacks were very lustrous, and they had faces of a dark or purplish red, and dark brown eyes.

The Wheaten Bantams are the cock breeding strains of Black-reds, Piles, and Duckwings. They are called Wheaten, from the resemblance in colour of the hens to wheat. There are Red Wheatens, and Grey Wheatens or White-necked Wheatens, and Pile Wheatens.

In the old fighting strains of Black-red Game fowls, the hens were generally wheaten or cinnamon colour. So also in Malays, and so in Game Bantams, and the cocks of these Red Wheatens are generally the most perfect in colour of all Black-reds. In fact, at one time we hardly ever saw a really good, bright coloured Black-red cock that was not Wheaten bred at least on one side, if not both. And so it was with the Duckwing cocks; unless they were "Wheaten bred" they had black stripes in their hackles, and lacked the bright, lustrous, and perfect colours.

The Wheaten cocks, then, are either Black-reds, Piles, or Duckwings. The hens we will now describe. Red Wheaten hen: Beak, greenish horn; legs and feet, willow or olive green; face and comb, red; eyes, red; head and neck feathers, golden, and as free from black as possible; breast, pale fawn or cream colour; thighs and underparts, creamy or pale buff (but the fluff is dark underneath); back and wings, pale cinnamon or wheat colour; tail, chiefly black, and the blacker the better, but the top outer feathers are generally edged with wheaten colour like the body colour.

The Duckwing Wheaten hens are like the above, except that their neck hackles are white, or white slightly striped; but the freer from black or dark coloured stripes, the better.

Pile Wheaten hens are of a nice cinnamon body colour, with the hackles showing rather more gold; the under fluff is white, and the flight feathers are white where the Red Wheatens show black; and the same with the tail feathers. In fact, the only difference between Pile Wheatens and Red Wheatens is the substitution of white for black, and in Pile Wheatens, yellow legs are preferable to willow.

The Wheaten hens we had more than thirty years ago were as good in colour as those we have to-day, though style has improved immensely, and we now have them much smaller.

Blue-duns and Blues have been produced in Bantam size by crossing Black-red or Wheaten Bantam hens with Blue or Blue-dun Game cocks, and we have had this colour three or four times. They are not very attractive in colour, and are not worth breeding for the show pen.

We must now pass on to consider how to mate the birds so as to produce the best results as to colour, etc., and will begin with Black-breasted Reds.

Many people say that it shows bad breeding on the side of one parent, if not both, when equally good cockerels and pullets cannot be bred from one pair of old birds. On this point opinions differ considerably we know, but without contradicting such an assertion, we must point to the results obtained from mating, specially for cock breeding on the one hand, and for pullet breeding on the other, and undoubtedly these results show that higher class and more perfect specimens are produced, and in greater average proportion, when birds are suitably mated to produce winners of one sex only. To give a recent illustration, we may instance the Black-red Bantam classes at the Crystal Palace Show, 1889, and there we find the Rev. Fred Cooper shows eleven cockerels in the open class of thirty-five entries, winning cup, and challenge cup, and first, second, third, and fourth prizes, and winning them easily, with birds which out-distanced all others a long way in richness and purity of colour, and these were especially of a cock breeding strain, with just enough Wheaten blood to keep the colour perfect, and to enable Mr. Cooper to head the lists at nearly every show he sent to, the season through.

On the other hand, there are the Messrs. Ainscough, who have bred, quite pure, the finest pullet breeding strain in existence, and at this same Palace Show they win first, second, and third with old Black-red hens, and first and challenge cup, and second, third, fourth, and fifth with pullets, just as easily as the Rev. Mr. Cooper won with cockerels; but Messrs. Ainscough, with ten cockerels (rare good birds, too, of the pullet strain) could only secure high commendation, because they lacked the lovely colour and bloom of the cockerel strain. Had the point of colour been ignored in the judging, we think two or three of Messrs. Ainscough's cockerels would have been placed high up in the prize list, as in shape and style, and shortness of feather, they were second to none.

We could give many more cases as conclusively proving our point, but think it unnecessary, and will, therefore, pass on to describe how we should mate the birds. First, for cock breeding, we should choose a very bright coloured Black-red cock, with good, sound coloured willow legs, not under a year old, nor over three years, and he must excel in every point. We wish to see in our cockerels fine, long, lean head; smooth skin on face and throat; bold, prominent, fiery eyes; upright carriage, short wings, long legs, good sound feet, broad shoulders, short back, fine, narrow stern, and close whip-tail; and with him mate two or three, or at most four hens (two-year-old hens are better than pullets for cock breeding). These hens need not be full Wheaten, for such are generally of too light a willow in legs to match well with the pullet strain, when pairs are shown together, but if they are half Wheaten-reds, we probably may find it quite sufficient. The hackles of these hens must be especially noted, as they are a sure guide as to their value, provided style and size, and length and fineness of limb are correct; and they should be composed of very narrow, short, close-fitting feathers, of a light, but bright golden shade, with hardly any black stripes visible; the back and wing may be either plain wheat colour, or show slight Partridge marking (we prefer the latter); the breast very pale creamy white, or the palest salmon colour; eyes, fiery red; ear lobes, as red as possible. From this pen we should confidently look for the very best coloured Black-red cockerels that can be bred, and pullets similar in colour to their mothers. Now, in order to breed the most perfect pullets for showing, it is absolutely necessary that we obtain a cock of the pullet breeding strain (and this no one can positively tell from his appearance, when an adult cock), and mate him with perfect-coloured hens. These cocks are generally, though not always, of a darker red, and of a more uniform colour (not so bright in hackle and saddle as the cock breeders), and darker willow or olive green in legs.

If possible, it is best to select one that has not been dubbed, for one can then see whether he has a small, neat and well serrated comb, and good red ear lobes; or to what extent they are defective. These points require more careful consideration in the selection of pullet breeders than in the cock breeders; as if, in the latter, they are faulty, all the faults are removed in the dubbing. A plan which, for very many years, we have adopted, is to select a number of our pullet strain cockerels, when about eight to ten weeks old, giving the preference to those whose feathers on wing and back are nearest to the correct shade and marking of the pullets, and from these choose out, when they have matured, the best birds to retain as future pullet breeders; and if our readers will adopt the same plan, they will find it the safest guide they can follow. It will be found that, to some considerable extent, we can influence the relative proportion of pullets and cockerels, according to the ages and numbers of birds bred from. Thus, when most cockerels are wanted, a small number only of hens should be in the breeding pens, say two, three, or at most four, and they should not be less than two years old, and quite in their prime. On the other hand, if a large proportion of pullets are wished for, mate an old cock (four or five years) with a large number (six or eight) of pullets, and the result will often be as high as 80 to 90 per cent. of pullets in the produce. In our own experience we have proved this over and over again.

But we must not lose sight of the fact that a great number of Bantam keepers and breeders cannot afford the room, nor the time and trouble involved by breeding from separate pens for cockerels and pullets, and, therefore, we will give our advice as to breeding when only one pen is available. In such a case, we should try to select as bright and good coloured a cock as possible of the pullet breeding strain, and put him with two very light golden hackled Wheaten, or half-Wheaten hens to breed cockerels from, and two of the best coloured pullets possible to obtain for pullet breeding, and early in the season, at feeding time, we should give more stimulating food to the Wheatens, to induce them to lay earlier than the others; in fact, give all the tit-bits to the Wheatens. When the chickens are eight to ten weeks old, mark all the cockerels from the Wheaten hens, so as to distinguish them and not breed from them, though they may be great winners; we say not breed from them—*i.e.*, if we want to breed show pullets—because their pullets will, as a rule, be foxey or rusty on the wing.

When sufficient numbers of chickens are hatched from the Wheaten hens, remove them—*i.e.*, the Wheaten hens—from the breeding pen, and you will then be certain as to the later hatched cockerels being pullet breeders. It is better to hatch the Wheaten-bred cockerels a month or six weeks before our show pullets, as the cockerels take longer to feather than pullets do.

If this be considered to involve too much trouble, it will be best to run only the Wheaten hens with the cock at first, and in this way there will be no fear of the two strains getting mixed, and so spoilt.

Brown-reds of the standard colour now breed very true to feather, and often we find own brothers and sisters heading their own classes at shows.

Pullets *can* generally be bred up to the standard colour from the highest type of winning cockerels and pullets; but there is a tendency in the cockerels to fall short of the desired extent of colour on the back and wing-bow. It will come all right after their first moulting season, but for many fanciers that will be too late. It is better, therefore, *for cock breeding*, to select hens that have the desired pale straw coloured hackles, well laced breasts, and *evenly-laced backs and wing-bows*, the feathers to be as rich green-black as possible, but their edges distinctly laced with pale gold or lemon colour.

Brown-reds cannot be improved in colour by crossing with any other colour of Game, except Birchens, but as new blood is sometimes needed, there has sometimes been a cross with either Black-reds or Duckwings. The results of either cross show plainly to a practised eye; and it takes several times breeding in, and back to the Brown-reds, before any dependence can be placed on their breeding true. The Birchen-greys are the result of crossing Brown-reds and Silver Duck-wings, and the surest way to produce good Birchen-greys is to mate a perfect-coloured

Brown-red cock with a very pale-breasted and silvery-backed Silver Duckwing hen. In the first cross there will be some very perfect-coloured Birchen-grey cockerels, and possibly some rich coloured Brown-red cockerels; and the pullets will be chiefly Birchen-greys, with too much lacing on the feathers of wing and back; any other than Birchen-grey pullets had best be killed, or at any rate not bred from. The only advantage gained by Brown-reds from this cross (Birchens) is in a paler lemon shade of colour where lemon is desired; but we are quite sure that this advantage is very heavily discounted, if not more than counterbalanced, by the longer and softer feather, which always accompanies it, and generally also by a loss in the darkness of the eyes.

Birchen-grey hens, of the first cross from Brown-red cock and Silver Duckwing hen, must be bred back to a good Brown-red cock, and their Birchen pullets again bred back to a sound Brown-red cock, in order to produce any large proportion of perfectly black wings, back and tail, with good grey hackle, and silvery lacing on the breasts of the Birchen pullets. The cockerels of this cross, while sounder in the black, will many of them show straw or lemon colour where they should be white; if, however, their top colour is too straw coloured for showing, their chief value lies in their being fit for breeding good Birchen-grey pullets from the show Birchen-grey hens, and also breeding the fashionable lemon Brown-red cockerels.

We wish it to be understood, however, that we are not much in favour of Birchen bred *Brown-reds*, because they are too "feathery." On the other hand, we greatly admire a handsomely marked Birchen-grey cock or hen *for its own beauty*, which we think fully equals that of a similarly marked Brown-red.

It may be gathered from the foregoing remarks, that when the Birchen cockerels come with too much straw or lemon shade, it is necessary to revert to the first or the second cross from the Silver Duckwing to recover the pure white top colour. And, we may add, considerable scope for skill, in breeding back to one side or the other, presents itself in the production of winning Birchen-greys, because we are almost certain to have lacing on the wings and backs of the pullets, or most of them, whose brothers are correct in colour and lacing.

Duckwings must now, for a short time, claim our attention; and in order that we may best understand how to proceed in mating our birds together, to breed perfect specimens, let us look back a few years and see what the birds have sprung from, and how the beautiful winners now seen have been produced.

About thirty-five years ago, we now and then could meet with a Silver Duckwing, but very seldom with a Golden, in Bantams. Our first was a Silver cock, which we mated with Black-red pullets. From these we bred several good-coloured Duckwing pullets, about twice the size of our Game Bantams of to-day, but all the cockerels had heavily striped hackles, and none had clear gold backs. The cross of Wheatens eventually produced the desired clear colour in the cockerels, but the two strains—*i.e.*, the pullet strain and the cockerel strain—had to be kept quite distinct, as one spoiled the other whenever they were interbred.

At the present day, however, through the more frequent use of the very best Black-reds of the pullet strain, we have strains of Duckwings that for many years have been so carefully bred for colour, that a large percentage of their chickens may confidently be looked for to come true to feathers, especially the pullets. There are, on the other hand, clever and skilled breeders, who, in order to breed their prize Duckwing cockerels, select a bright coloured, clear hackled Black-red cock to mate with their Duckwing hens, and generally get a few good Duckwing cockerels with clear creamy hackles. All the pullets bred in this way are certain to be Reds or Wheatens, but no Duckwings. These pullets are in turn useful to breed with Duckwing cocks again.

Much care must be taken to avoid breeding from those Black-red hens that have dark caps—*i.e.*, dark feathers on the top of the head, where they should be golden—for the Duckwing pullets bred from these hens are sure to have brown caps, which look worse on grey heads than on golden ones, inasmuch as the contrast is greater; and further, the Duckwing pullets bred from such hens almost always fail in colour of back and wings, though usually they have beautiful salmon breasts. As Golden Duckwings, then, are the result of crosses between Silver Duckwings and Black-reds and Wheatens for the cockerel breeding strains; and for the pullet breeders, the crosses between Silver Duckwings and the pullet breeding strains of Black-reds, it has often proved a very difficult task to get these constituents in exactly the right proportions, and much skill and judgment is required, even now, when introducing fresh blood, that there be not too much of the Black-red, or the produce will be chiefly of that colour. We are even more particular in choosing our pullet breeding Duckwing cockerels when in their chicken feathers, than our Black-reds; there is not only the same fineness of marking to look for, and the same fineness of head points, neat and perfectly straight comb, red lobes, etc., but there is the silveryness of the feathers on the back and wing, which is lost in the change to adult feather. This silvery sheen is a very important point to note, and may be taken as a sure guide as to the colour we must look for in the pullets bred from its possessor.

Remember that we want our show Duckwing pullets to have a fine silvery margin to all their wing and back feathers, and we can best ensure this by breeding from a Duckwing cock which showed the same point, in a marked degree, while he was a chicken.

We think that now it is easier than, perhaps, ever before, to find the best materials for Duckwing cockerel breeding, and we prefer to use a very bright coloured, clear hackled, Black-red cock, mated with pale-breasted Duckwing hens, with just a little Wheaten blood in them. Of course it would be perfectly useless to expect a single Duckwing pullet from such a pen, but there should be plenty of rich coloured Duckwing cockerels, and Black-red cockerels, and cockerel breeding pullets; some of these latter might be useful to mate with a good-coloured Duckwing cockerel or cock for another season. We have said nothing yet about the breeding of Silver Duckwings, but it will be necessary to note that there are different strains of Silver Duckwings, as of Goldens—*i.e.*, there are Silver Wheaten hens, the cockerels from which are just the same colour (except, perhaps, a clearer white hackle), as the cockerels from the Silver Duckwing, or, more strictly speaking, Silver-grey hens.

At the present time, we can place our hands upon perfectly reliable strains of each of these varieties. The Silver Wheaten strain is useful, occasionally, to cross with Black-reds to produce Golden Duckwing cockerels. The Silver-grey Duckwing strain is the foundation of the show Duckwing pullets, both Golden and Silver, and this strain can be, and has been, in-bred for very many years without deterioration.

Just now we observed that the Silver Wheaten strain of Duckwing cocks have clearer hackles than the other Silvers, and very often the hackles are perfectly clear silvery-white. Such cocks, when mated with the cock breeding strains of Black-reds, are perfectly reliable for breeding Golden Duckwing cockerels and Wheaten pullets, but if they were mated with the show hens of Duckwing variety, nothing but disappointment might result. It is, therefore, necessary to know the origin and ancestry of the stock birds bred from. The Silver-grey Duckwing cockerels, when half grown chickens, are of exactly the same colour and markings as the pullets, and these are always reliable pullet breeders, when mated with exhibition Black-red, or Duckwing hens or pullets.

Red Piles are not so difficult to breed up to the required standard as are the Duckwings. They were originally produced by crossing the Black-red and pure White Bantams, and for years large numbers of the chicks used to revert to one side or the other; but now it is a very rare occurrence for a pure White chicken to be bred from Piles, though far too many are so pale in colour as to be called Lemon Piles or White Piles. In such cases where the light chickens are the major portion, it is best to select a sound-coloured Black-red cock, and mate him with the Lemon Pile hens, especially the yellow legged ones, and the result, nearly always, is a good majority of rich Red Pile cockerels, but too many Black-red pullets; these can be bred back to a Pile cock of good colour, with good results. We have found that once in five years is quite often enough to revert to the Black-red cross, in order to keep up sufficient colour in our Piles.

Even in this breed, it has been found desirable to maintain two distinct strains, one for cockerel breeding and the other for pullets, and it will readily be seen why—for the same rules hold good in this colour as in the other varieties, Black-reds, Brown-reds, and Duckwings, viz., that the best pullet breeders hardly ever produce many cockerels with sufficient colour on the wing-bow and saddle. The best cockerel breeding Pile hens usually are "high-coloured," or "rosed," on the wing, instead of being milk white, and they are also laced with gold on the back, and have more golden hackles. Some very excellent Pile cockerels are bred between Pile cocks and Red-wheaten hens. These cockerels have, as a rule, very white breasts, and good white where white is wanted. Occasionally a few Pile-wheaten pullets are produced, and they should certainly be retained as cock breeders, for which purpose they are invaluable.

Yellow legs (as deep an orange yellow as possible) are the favourites in Piles; but if yellow legs only are bred from, in a few years the colour decreases, and nearly white legs become the rule, rather than the exception. It is better, therefore, to cross yellow-legged with willow-legged birds, and by this means we keep up the bright orange legs. About once in three years is often enough to cross the yellows and willows. In breeding Piles, special care should be taken to secure very hard-feathered stock birds, especially the cocks, for all the lighter coloured breeds are apt to become soft-feathered. Generally, a good hard-feathered Black-red cock is the best bird to use in order to harden and tighten the feather in Piles. If, however, in thus crossing with Black-reds many of the chickens show distinct bars of stone colour on the white ground, where the feather should be pure white, then it is proof that there is quite sufficient Black-red blood in the Piles, and they should be either bred in-and-in, or else bred back to Piles only.

And now we must consider how to breed the best show pullets in the Pile Bantams. Our plan would be as follows—viz., to select two or three of the very best winning pullets we could find—*i.e.*, the clearest white on wing, tail, and back, with rich, deep salmon breasts, and nice golden-laced hackles, rich orange legs, or sound willow if yellow cannot be got—and with these mate an old cock, say three or four years, if a healthy and vigorous bird, deep chestnut on the wing ends, and clear in the wing-bar; if lightly laced on the breast at this age, it need not be regarded as a fault in the pullet breeding pen, but rather otherwise. If the produce come out rather high-coloured on the wings of the pullets, select the whitest winged, and whitest breasted of the cockerels, if he has enough colour on the ends of the flights; and if he has very little colour on the shoulder or wing-bow, all the better for the pullet breeding, and next season let him occupy the place of his sire. You are then almost certain to obtain good pullets. We have often bred very sound-coloured Pile Pullets by mating a pale-coloured Pile cock with our best pullet strain Black-red hens. The worst fault of the Piles so bred, is that they nearly invariably have dark—*i.e.*, willow legs—but in feather they are irreproachable, and if judiciously mated back to a good yellow-legged cock, plenty of the same coloured legs may be expected in their chickens.

The same difficulty arises here as in the other colours, as to which cocks will breed the best pullets—*i.e.*, when you are choosing from old cocks.

In the case of cockerels half grown or three-parts feathered, it is best to choose those which are the clearest and purest in the white, and most distinct bay on the wing ends, and with only a moderate amount of colour on the wing-bow and saddle.

Black Game Bantams are very nearly, if not quite, extinct; we have not seen one worth mentioning for several years; but, as we stated on a former page, we once produced a strain of very lustrous Blacks from the same stock from which we produced White Game Bantams and Brown-reds—*i.e.*, all can be traced back to a Black Game hen which we added to our Brown-red breeding pen in 1865, for the purpose of improving shape, and lustre, and hardness of feather. If anyone inquired of us how they could now obtain Black Game Bantams, and how long it should take to produce them, we should say the time required would probably be about four years; and our own plan of procedure would be to begin by obtaining the smallest pure Black Aseel cock we could purchase or borrow, and run a couple of White Game Bantam hens with him, from which we should endeavour to raise three or four broods, and probably have several Black pullets, but perhaps not one Black cockerel; nearly all cockerels would be coloured. Then for the next cross, we should run the Black pullets bred from them, with the Black Aseel cock, and this time there might probably be some Black cockerels that would only have a few red or straw feathers in hackles or saddles, and possibly there might be a perfectly Black cockerel, as a case of reversion to the Black hen of 1865; this would not at all surprise us. But whether or not there were a solid black cockerel, we should select the *blackest* grandson available from the Aseel cock, irrespective of whether single or triple comb, and breed him and the blackest first-cross daughters of the Aseel cock together, and the next season, the produce of these back to their sire. Probably the average weight of the birds now would be just about 2 lbs. for the cocks, and a trifle under for the hens, when in hard flesh and condition; and there would be plenty of pure black hens and pullets, but few cockerels entirely free from colour in hackle and saddle. We should then, to secure fixity of Game Bantam character, single combs, etc., run the blackest cockerel last bred, with the pure White Game Bantam hens, and some of their produce would be certain to come either all black, or, what is more probable, a mixture of the pure black and Brown-red—*i.e.*, the cockerels would have some coloured feathers in hackle, saddle, and wing-bow, and the pullets would be entirely black with the exception of a little colour in the hackle. The average size of these would hardly exceed ordinary Game Bantams, say $1\frac{3}{4}$ lb. for full-grown cockerels, and $1\frac{1}{2}$ lb. for pullets. The strain would then be so far made, that only a judicious selection and in-breeding between the above families would be required to perfect the Black. Aseel, or triple combs, would keep cropping up, now and then, but would be in a minority, and increasingly so each season, if they were not bred from. There would be no difficulty experienced as regards the colour of the legs, for, with very little variation, they would be the darkest olive or nearly black, just showing a little yellow under the feet and between the toes. This would be our plan. But there is another way, and that is to select a small Black Game hen or pullet, and run with her a pure White Game Bantam cock; but if we tried this plan, we should not trust to one hen only, we should have either two or three, as in all probability many eggs would prove clear. When once the first-cross chickens are grown up, it would be necessary to select the blackest cockerel to mate with the White cock's own sisters or relatives—pure White Game Bantam hens—and mate the White cock with his own Black pullets, and then breed between the two families. By this means you would gain the advantage of certain fixity of single combs, but you would as certainly lose hardness of feather, and fine whip tails, and the bright red face and eyes now required in Black Game.

Many times within the last three or four years, we have been asked, "Are there any Blue or Blue-dun Game Bantams now?" We fear not, but there may be; for, the last time we stayed a few days in Edinburgh, it was at the house of a gentleman in Grange Road, who had a few of our old Blue-duns, and was trying, from them and a pure Blue Old English Game cock, to raise a strain of pure Blues of Bantam size.

If any of our readers are so inclined as to make a strain of their own, it can easily be done now by procuring a sound Blue Game cock, the smaller the better, and breeding with him and pale Lemon Pile Bantam hens; the very first cross might probably throw Ginger-reds, Piles, and Blue-duns, and by selecting each year the smallest birds of best blue colour, the strain could easily be produced and maintained.

There were, about twenty-five to thirty years ago, many very pretty Spangled Game Bantams bred in the neighbourhood of Southwell, Edingley and Farnsfield, Notts. They were chiefly Red-spangled, though there were also some Greys. In each case the hens were exactly as if Black-red or Duckwing hens had been out in a snowstorm, and were sprinkled over with small snowflakes. The Red-spangled cocks were like Black-reds, except that they were spangled evenly with white all over, and the tail and flight feathers were about equally white and black, and the flight coverts white and bay. The Grey-spangled cocks were like Duck-wings, only well spangled with white. All these, both Greys and Reds, had yellow legs, or the lightest shade of yellowish-willow, and the chief breeders of them were the two brothers Henry and Frederick Schumach, the latter, along with his family, still residing at Southwell, though Henry has been dead several years. These colours are frequently seen in Old English Game, and also in Aseels, and we used to have them in Aseel Bantams, of which more anon.

CHAPTER VI.

MALAY, INDIAN GAME, AND ASEEL BANTAMS.

THESE three varieties are the most closely allied to the Game Bantams, to a large extent resembling them in shape, habits, and hardness of feather, and yet the distinguishing points are quite well defined, as will be noted the further we proceed with the description of each.

In passing, we may here remark that the writer of these pages, after years of labour, was the original producer of all the varieties of Malay Bantams, Indian Game, and Aseel Bantams, and that, to the best of his knowledge, all these varieties now extant have sprung from the birds he bred down from the large varieties, of which they are the miniatures. To begin, then, with the Malays: These are now bred of five different colours, viz., White, Pile, Dark Red, Bright Red, and Pheasant, and of these the Pheasant and the White are chief favourites. In shape they are the exact counterparts of the large Malays, excepting, what will naturally be considered almost inevitable, that they are hardly so massive in the eyebrows proportionately to their size. Though only slightly larger than Game Bantams, they are stouter built, stand much taller, and have broader shoulders; narrow sterns, drooping tails, with very narrow sickles, and very many more of the narrowest and most wiry side sickles than any Game Bantams should have. They all have orange yellow legs and bills, whatever the colour of their feather; and they all have the strawberry or half walnut comb, and white, pearl, yellow, or daw eyes. The three drooping curves of the large Malays are exactly reproduced in these Bantams, and altogether they have been very highly spoken of by the London Press from time to time. The first colours to make their appearance were the Black-reds, dark or light, or Dark-reds and Bright-reds.

These were followed a couple of years later by pure Whites, and then by Piles; another two or three years passed before the Pheasant Malays were shown, and they came out at the Crystal Palace, where, like the other colours, in turn before them, they were successful in winning the Cup over three or four classes.

Before proceeding with the directions for breeding these colours, we will just point out a few advantages they possess over their Game rivals.

First, while fully equalling the Game Bantams in beauty of colour of feather, their bright yellow legs make them more striking looking; their combs never give any trouble by growing out of shape, and they never want dubbing; they never want their ear lobes removing, as they are always red; the only preparation required for the show pen is to have their combs, faces, legs and feet clean washed, and, of course, the birds themselves to be tame and used to a pen. They are as hardy again as Game Bantams; their eggs nearly always hatch, seldom a clear one, and from the first the chicks are strong and vigorous, and are fit to show at five months old. They make excellent parents, and bring up their own chickens till they are quite able to take care of themselves.

We have known one pen of them, that we sent to run in a park, roost all through one winter in the top of a lofty Ilex tree, and they kept perfectly healthy and hearty.

The usual number of eggs laid by a Malay Bantam hen is ten or eleven, and then she wants to sit, *i.e.*, if undisturbed, and she will sit three times, and bring up three broods in one season, if allowed. The eggs are large for the size of the bird, and the smaller end pointed.

And now with regard to colour. The Black-reds are of two shades of colour, and, in fact, two strains. The Bright-red cock and Partridge hens are of one strain, and in colour resemble the Black-breasted Red Game Bantams, only that they are rather darker, and the hens have much darker hackles and heads. This strain will reproduce with great regularity, both cockerels and pullets. The Dark-red cock has deep red hackle and saddle, dark purplish crimson or maroon back, and wing-bow, and all the rest of the feathers raven black, and very lustrous. The hen's hackle is dark bay, and her body, wings and breast are wheaten or cinnamon—the more even in shade all over the better. These vary from a light wheaten to a warm, rich cinnamon shade of colour. This strain also breeds true to colour. The Pheasant Malay is in colour exactly like the Indian Game, *i.e.*, the cock has raven-black hackle, breast, tail and underparts; dark maroon wing-bow, and the same colour and black on back and saddle. The flight coverts good rich bay, in many of the best birds well laced, or lined with black on the edge of each feather, and this is a point to be coveted. Such cocks always breed the best laced pullets. The colour of the hen is as follows:—Head and neck bluish-black; breast, wing, back and thighs, one uniform rich bay, every feather distinctly laced with the same shade as the neck; in some it is blue-black, in others dark beetle-green. As in the large breed from which we bred these down, we find some pullets double laced, *i.e.*, one lacing inside the other, but more single laced, and this is the more showy of the two styles of marking.

When these Pheasant Malays are newly hatched chickens, they are all quite light buff; but when ten days old, and their first flight feathers commence to show themselves, the regular and distinct marking at once is manifest, and as they grow up they become often quite black in the breast, and it is not until they are eighteen or twenty weeks old or so, that the perfect lacing of the full feather shows in all its beauty. This is really a charming variety to keep, and one that a fancier never tires of.

Piles need no special description as to colour, for in that respect they are exactly like Pile Game Bantams.

Whites, too, will need very little description, save only that we would caution our readers against crossing Whites with any other colour, and above all not to use a sandy-backed cock to breed from, or they will regret it as long as they keep the produce, with the hope of breeding any pure Whites from them.

We have received several letters asking us to say how we first crossed our birds to obtain the Malays, and we think we must, for a full answer, refer them to a prize essay of the writer of these pages, published in "Fowls for Pleasure, Prizes, and Profit," March 13th, 1890, which was written in response to an invitation published previously, offering a prize for the best essay on "How to breed Malay Bantams." From this essay we give two or three extracts, viz., "This question puzzled me many years ago, at a time when there were no Malay Bantams, nor anything nearer in type to the Malays than Game Bantams. I had, for years, bred all colours of Game Bantams, and one day I resolved that I would breed Malay Bantams, so I tried in various ways to obtain the first cross between large Malays and my Bantams, but for a long time I met with nothing but disappointments. At last, however, I found that one hen's eggs were all fertile, and I had them set in a house all by themselves, and reared a fine brood of chickens from them; and from that brood all the Malay Bantams, both at home and abroad, have sprung. I may here mention that the only breeds used in the production of my Malay Bantams have been pure bred Malays, Indian Game, Aseels, and Game Bantams." "In answering the question, 'How to breed Malay Bantams,' I would say that the easiest and simplest way would be to buy the best pen that can be had (they are not so costly as some breeds of Bantams are), and give them a small house and nice little run," etc.

"A still less expensive plan is to buy a sitting or two of eggs, and thus for a guinea, or a couple of guineas, a first-class stock can be raised. But if anyone put the question to me, 'How would you advise me to start now, to breed down from large Malays, and make a strain of Malay Bantams of my own?' I would answer: If you have time, and patience to do it, you will in the end obtain the best results in this way:—First select a highly-bred, typical headed Malay cock, as short feathered as possible, and hard and wiry looking in tail, etc., and having broad, flat skull, heavy overhanging eye-brows, and a very firm and flat comb. Mate such a cock with a couple (not more, or else they will kill each other) of the smallest Aseel hens you can get, each having a broad, flat skull, heavy eyebrows, sunken eyes of the clearest pearly whiteness, and low carried, close tail, that moves from side to side in a nervous manner with every step taken.

"From these, you will obtain some cockerels and pullets with correct Malay type of head, strawberry comb, etc. Select the smallest of these cockerels that has Malay character, especially in head, and mate him with one or both of the Aseel hens when he is six months old, and breed at once. The smallest cockerels from this mating (the second cross) should be small enough (about three pounds at seven or eight months old) to breed with Game Bantam or Malay Bantam hens, and the pullets with Malay Bantam cocks. And now by careful breeding, etc., and always choosing the smallest chickens of Malay character to breed from, you will obtain show winners." The other portions of the essay omitted here will be found worth reading, and if the number (it is No. 140 of Vol. III. of "Fowls") can be obtained, we advise our readers to buy it, but if it is out of print, we will supply the missing parts to those applying to us for them.

We next come to consider the Indian Game Bantams, of which there are but few as yet, and they are the exact counterparts, in miniature, of such birds as the Messrs. John and James Frayn's noted cup winners at the large shows; but whereas the large birds weigh from 7 to 12 lbs., these little beauties weigh from 1¼ to 2 lbs.

In shape they are less tall and reachy, but more compact, than the Pheasant Malays, and their heads are not so broad, the eyebrows less deep and prominent, the eyes similar in colour, *i.e.*, pearly white or straw colour; the comb is always triple and very neat, like the best Aseels and Brahmas, the tails less drooping than the Malays, but still carried low and close, the backs shorter, and the bodies rounder and plumper; in other points they are almost exactly alike, and as for mating them for breeding, the rules are exactly alike, and need not be repeated.

We have had a few, now and then, pure milk white chickens from our coloured Indian Games, evidently cases of reversion, for very many years to the days when we spent much time and trouble over making Aseel Bantams. Of these we had Black-reds, Blacks, Whites, Piles, Spangles, Pheasants, and Greys.

We must confess that though we have bred hundreds of Aseel Bantams, and won a great number of prizes with them, and really consider them both elegant and useful, considered as a distinct variety; yet when we bred them, it was not with the intention of always keeping them as a distinct breed, but rather because we had two other ideals in view, in the production of which we considered our Aseel Bantams the chief foundation stone, as we may say, namely, the Indian Game Bantams and Malay Bantams.

We have now in one long grass run a pen of cock and seven hens, pure white Aseels, the cock about 22 ounces, and hens 18 to 20 ounces each, and for elegance and symmetry they will be hard to equal. When we find our White Malays show a tendency to grow too large, a cross with these Aseels soon brings size down again, and still they are hardy, plump, quick, and agile little birds. The two colours we found most useful of all were the Whites and the Black-reds, the hens of the latter being Cinnamon.

Besides these we had pure Blacks, which were as lustrous as Black Hamburghs, and have been useful to us in imparting glossiness to our Pheasants. But the colour with which we won the most prizes was the Red Spangle, a very showy little bird, and quite distinct from any other Bantams exhibited about the same time. Most of our Aseels came from Devon and Cornwall, but one which we obtained from the Hon. and Rev. F. G. Dutton, a beautiful little Cinnamon, proved the most useful to us in the production of our Red Malays; and some of Mr. Bryan's birds helped us equally well with the Whites.

The Aseel, being a bird of very great antiquity, possesses such a prepotency, when crossed with almost any other breed, that, with very little trouble, it stamps its outward characteristics upon the offspring of any breed with which it is crossed. It almost seems to swallow up any other variety's individuality. On this subject we related some of our experiences in a paper we contributed to the 1890 Christmas number of the *Fanciers' Gazette*, and to which we may refer shortly, when we come to describe the Brahmas and Partridge Cochin Bantams, and how we produced them.

It may probably have been observed by some readers that when we were describing the points we should seek for, in the Aseel hens, to mate with Malay cock, in originating a new strain of Malay Bantams, we spoke of Aseel hens with *broad, flat skulls, heavy eyebrows*, and *sunken eyes*, not because all those points are desirable in an Aseel—though often enough seen in the breed—but because such Aseels would be of greater value for Malay breeding, than would the better exhibition Aseels prove, whose eyes should be bold and rather prominent, though not so much so as English Game.

WHITE MALAY BANTAMS.

Produced by Mr. W. F. Entwisle, and the Property of Miss E. H. Entwisle, Calder Grove House, near Wakefield.

Cock, never exhibited. Hen, winner of First Prize at Dairy Show, 1893.

Before closing this chapter, we may be pardoned, perhaps, for saying that the breeds described in it are eminently suited for beginners in poultry keeping to commence with, on account of their hardihood, and that there are less complications in the breeding of both sexes to the standard requirements ; and further, there is not yet such keen competition, in shows, in these classes as will be found in many of the older breeds.

CHAPTER VII.

PEKIN OR COCHIN, BRAHMA, BOOTED, SULTAN, BURMESE, AND SILKY BANTAMS.

THE Cochin or Pekin Bantam is evidently of very ancient origin, and to the Chinese belongs the credit of their production, and cultivation, probably for many ages.

Their first introduction into England was not until the year 1860 or 1861, and previous to that, we believe no fancier in this country had ever heard of them. These were all Buffs—the cocks rather a rich darkish cinnamon, and the hens some shades lighter. Several years elapsed before Cochin Bantams of any other colour were heard of (and by that time the Buffs had almost died out, and were only in the hands of one breeder). These new ones were Black, brought over by an officer from China; the hens were excellent in every particular, but the cocks were none of them free from white, or straw colour in the hackles, and several had brassy wings. However, careful breeding has changed that, and now we find a few that scarcely show any white, even in the under fluff of the hackle.

Again some years elapsed without any other colour of Pekin Bantams being heard of as in existence, and, meanwhile, the writer of these pages was trying his best to produce a lighter shade of Buff cocks—more like the Lemon Buffs so fashionable in the large Cochins at this time—and also sounder coloured Blacks, for which purpose, as well as to give tone and vigour to the constitution, he introduced a cross of the purest and best White-booted Bantam he could find, also intending, eventually, to breed White Cochin Bantams. While these were gradually developing, he produced the first barred specimen, with the cuckoo marking very pale and indistinct, but which became the foundation of all the present Cuckoo Pekins, though strengthened, and aided by the kindness of Mr. Leno, who has the credit of importing from China the first and only Cuckoo Pekin cock that has reached our shores as yet; but of this more anon.

The cross of one of the earliest Black cocks we bred with Buff hens produced some cockerels so near the colour of Partridge Cochin cocks, and at the same time so perfect in shape, symmetry, and size, that we resolved upon making, and so far as we could, perfecting, a strain of Partridge Cochin Bantams, with which prizes have already been won at the Crystal Palace, and elsewhere. Meanwhile, Whites were improving rapidly, and now are quite a recognised variety, having won two cups at the Palace (besides cups at other shows), and many other prizes. So that at the present day, 1892, we have five distinct colours, or breeds of Cochin Bantams, seen in the prize lists of our best shows, viz., Buff, Black, Cuckoo, Partridge, and White.

We must now look at each of these in turn, and our endeavour will be to present the fullest information we possess, up to date.

To begin, then, with Buffs, as undoubtedly the earliest we are acquainted with. As we stated in our opening remarks, these came to us direct from China in or about the years 1860 or 1861. We think that it was in the year 1860 that the Emperor of China's Summer Palace in Peking, or Pekin, was captured by the allied British and French arms, and among the spoil then captured were some beautiful Buff Cochin Bantams, which were prized and cared for by a "British Fancier," who was an officer then faithfully serving his Queen and country in China. This officer sent them home to a friend, who bred a few, and then let Mr. Kerrick, a gentleman residing near Dorking, have them. These birds were bred in-and-in for nearly twenty years, and for a dozen of which they seemed to show no marked ill effects from the mode of in-breeding; gradually, however, the constitution became weakened, and sterility ensued. During

BUFF PEKIN OR COCHIN BANTAMS.

Bred by Mr. Entwisle, and the Property of Mrs. Entwisle, Calder Grove House, near Wakefield.

Cock, winner of First Prizes at Thorne, March, Lincoln, Market Harbro', 1892; First and Cup at Crystal Palace, First at Tunbridge Wells, etc., etc., 1893.

Hen, First Prize at Tunbridge Wells, 1893; Second at Otley, etc., etc., 1894.

these years, we believe we are correct in saying that the late Mr. Beldon purchased from Mr. Kerrick every chicken he would part with, and these found their way through Mr. Beldon's hands to, among others, Mr. H. B. Smith, of Preston; Mr. W. J. Cope, of Barnsley; Mr. John Newsome and Mr. J. S. Senior, of Batley, who were unable to make headway with them through their delicate constitutions.

Two of these gentlemen, Messrs. Cope and Smith, tried to strengthen and invigorate them by crossing with Nankin Bantams, and the chickens certainly were not spoilt in colour, but, as certainly, they lost character, and became long and weedy in body, leg, and tail; scantily feathered on shanks and feet, and many of them dark legged. Then, one day, we heard that Messrs. Bailey and Sons, of London, had received a fresh consignment of fresh Buffs from Shanghai, but of these we never saw a feather.

About the year 1884, when we had made considerable progress in breeding down from the large Buff Cochins, and had quite a nice lot of Buffs between 2½ and 3½ lb. weight, we were fortunate in importing a dozen good Buffs, which reached the London docks, and our own home next day, in safety; and in January, 1885, we divided this consignment into two equal lots, (Mr. E. Walton taking his choice), keeping the remainder ourselves, which greatly improved our own Buffs in size and shape, while our birds added vigour and stamina, the result being that now our Buffs are as hardy as any of our Bantams. One of our cock birds, a cup winner here, which we lent for the breeding season, 1890, to a well-known poultry judge in the U.S.A., after we had hatched thirty chickens from him, stood the voyage out well, had a successful season as a stock bird out there, moulted well, and returned in time for our winter shows, looking none the worse for his twice voyaging across the Atlantic. He is breeding well this season, and walking his quartette of hens across the lawn in front of our window as we are now writing. This is more than we could reasonably expect of any of the imported Chinese birds.

About the year 1885, we became aware of the fact that a few Buff Pekins were bred in the U.S. America, and at considerable trouble and expense, we secured some of the best birds that could be sent from that country. These we found superior in colour to our pure Chinese, but deficient in foot and shank feather, and quite wrong in shape; they were longer in neck and leg, and longer in back, and sloped from the shoulder down the back to the tail, which suddenly rose from a divided saddle. However, for the sake of the colour, we kept and bred from some of these Americans, and admit that they have been very helpful when bred with the better shaped and heavier cushioned birds we previously possessed.

One great point we value most highly, and we think our English breeders will not be long in recognising, is the sound, even colour insisted upon by the Americans. They say, "A Buff must be *buff*," perfectly free from any dark shade in fluff or flue of feather, buff under the wings when expanded, buff in all the tail feathers and foot feathers. A bronze tail is considered a blemish, and the Americans don't allow such faults to be hidden, or disguised by pulling out the faulty feathers. Why should we allow it?

If all judges resolutely set their faces against such practices, and never gave prizes to birds so exhibited, we may depend on it such birds would not appear at shows; people would buy or breed, and show the right sort.

At the present day we are breeding from four yards of Buffs, and not one bird has dark feathers in it. Others can do the same if they will. Buff Pekins, as a rule, are slightly larger than they were thirty years ago, but they make up for any increase in size by their increased usefulness. They are capital layers, sitters, and mothers, and most docile and affectionate in their disposition. A good Pekin cock always has the peculiar Cochin crow, with a long sustained finish, quite different from the crow of a Game Bantam, or a Rosecomb Bantam voice.

Pekins' eggs vary in size and colour, many being white, while others have the usual tint of the large Cochin eggs. Most of our Buffs lay tinted eggs—all our Partridge eggs are tinted; many of the Blacks, Whites, and Cuckoos lay white eggs. The usual number laid is nine or ten, and then the hens want to sit and hatch. We usually, now, run five with each cock or cockerel for the first two seasons, but only two or three hens when breeding from a cock over three years old. We make their nests on the ground, and only raise the perches sufficiently high for the birds to walk underneath without touching them.

Pekins are very short in wings and legs, and

inclined to be fat and heavy, so they are very poor flyers. A six feet high fence of wire netting is quite sufficient to keep them within bounds. They require a liberal supply of green food, and *no maize*, nor hemp seed, nor any fattening diet.

In mating Buff Pekins for colour, we cannot do better than follow the Chinese lead ; if we deviate from this, uncertainty at once meets us. Our opinion, after more than ten years' experience in breeding Buffs for colour, and some years with five or six experimental pens, is that a solid-coloured, rich, but not dark, Cinnamon cock, and a clear, even-shade Buff hen, two or three shades lighter than the cock's breast, should always be looked upon as a correctly-matched pair, whether for breeding, or for the show pen. If, on the other hand, we take for a breeding pair, a hen whose colour matches that of an ordinary coloured Buff cock—*i.e.*, a full, warm colour—instead of breeding cockerels the same colour, they will get darker and darker every year, until the cocks would be dark chestnut, and the hens dark cinnamon. If the reverse mode be adopted, and we select light Buff cocks and light Buff hens, both of one shade, the tendency is to breed still lighter chickens, and generally the cockerels have pale hackles and breasts, with wing-bow of a different colour, and often mealy, and the pullets are what is termed mealy or mottled. We have always had the best results from mating a full Cinnamon cock and moderately warm Buff hens, about two shades lighter in body colour than the cock's breast; and if our readers will follow this plan, and always avoid cocks or hens with any unsound colour, either shades of black or white, they won't get far wrong in colour.

Now as regards shape. This is of great importance, and as our remarks on this point will apply equally to all Cochin Bantams, whatever their colour, we shall not need, we think, to refer to this subject when describing the other varieties.

With regard to shape, the Pekin or Cochin Bantams should be exactly the same as the larger Cochins, whilst as regards size and weight, the latter point should not exceed one-fifth the average weight of Cochin fowls—*i.e.*, about 2 lbs. to $2\frac{1}{4}$ lbs. when full of flesh for the cocks, and $1\frac{3}{4}$ lbs. to 2 lbs. for hens. Of course, it is a very easy matter to bring the birds many ounces under these weights, but the above may be taken as correct for sound, healthy birds, well cared for, on their ordinary runs.

And now as to shape. The cock should be small and fine in head, though stout in beak, fine and neat in comb, which must be well serrated and single, and perfectly erect.

Pekins are, as a rule, much larger in comb—in fact, twice as large, in proportion to their size, as the Cochin fowls—but this is undesirable, and in a few years, no doubt, much neater and smaller combs will be the rule. The face must be quite smooth and fine in texture of the skin, and the ear lobes and wattles must be long, ample and smooth in texture, and, with the comb and face, of brilliant red ; the eyes should be as near red as possible, though generally found to be yellow, and sometimes pearly white, which is not to be desired. The neck should be short and full, and neatly arched ; the body carried slightly forward, and the top of the tail as high as the head ; the back short and broad, increasing in breadth to the saddle, which should be very full, and rise well from between the shoulders ; the wings very short, and tightly tucked up ; the tail abundantly furnished with side hangers, or side sickles, but having no hard quill feathers, the tail proper (as in the hen) being composed of very soft feathers, the quills of which are fine and thin, and unresisting to pressure ; the whole tail should fall nicely over in an unbroken curve with the back and saddle, the fluff underneath very full, and standing well out ; the thighs short and broad, and set well apart, and the shanks thick and short, and the whole leg and toes abundantly covered with soft feather. The hen must have very small head, with gentle expression ; comb, neat and small, well serrated and perfectly erect ; ear lobes and wattles, red, fine skinned and thin ; eye to match the cock's, red if possible, but if not red, then yellow comes next in value; neck, short; and hackle full and long; body feather very full, abundant, soft and fluffy ; wings very short, and well tucked up to the sides ; cushion full, and round as possible; tail very small, and nearly hidden by the full cushion ; legs very short, and with the feet and toes abundantly feathered. The body should almost touch the ground when the hen is walking, and give one the appearance of plumpness, and a very happy and contented disposition.

Cochin Bantams, and especially the cocks, do not acquire their full shape or complement of feathers during the first year. Many splendid two-year-old birds were but narrow as cockerels, and short of cushion, and an inexperienced fancier would probably have considered them of no value when half-grown birds. Then with the pullets, it is frequently the case that those which have the roundest and fullest cushion as hens, were quite inferior as regards cushion when young pullets, scarcely even showing any at all.

We are always most particular, in selecting a brood cock, to choose one as short and broad in the back, and especially as full and broad in saddle, as possible, and as short in legs, and heavily feathered on thighs and below hock joint. The cushion should rise from the base of the hackle, and form one unbroken curve up the back, and ending with the curve of the tail feathers. And the brood hens should show these same curves still more plainly, as their feathers are so much broader and rounder at the ends than are the cock's feathers. Particular care should be taken that the tail feathers, and all the hock feathers, are very soft in the quill, the feathers themselves having a disposition to curl inwards. These are sure signs of high breeding, when accompanied by abundant feathering elsewhere.

The legs and toes of all Buff Pekins (and we may say all Pekins, Blacks, perhaps, excepted) should be as rich a yellow as possible. The more abundant the shank and toe feathering, the better, some of the best birds having the inside of the legs as neatly feathered as grouse. Great stress should be laid upon ample feathering of the middle toe right up to the toe nail. This is difficult to obtain without "vulture hocks" (stiff feathers from the hock joint, sometimes touching the floor) being an accompaniment, which is very undesirable, though not now, as once, considered a disqualification. When a brood of young Pekins is hatched, we find little trouble in rearing them, according to directions given in a former chapter, and they grow and feather very rapidly, looking as pretty at seven or eight weeks old, as ever in their lives.

We now pass on to consider the next favourites —Blacks. To begin with the foundation (feet and legs), it is a disputed point whether Black Cochins should have yellow or dark legs. We decidedly prefer yellow, and next to yellow we should choose darkish or dusky legs, showing yellow under the foot and between the toes, and yellow under the scales on the shanks. Yellow is, however, scarce, and many breeders consider it an indication of a cross between Black and some other colour, possibly White.

However, as the exact colour is not insisted upon, we may let that pass, merely remarking that blue legs, or white legs are considered *very bad*.

In Black Pekin Bantams, colour of feather, and brilliancy of sheen very properly count highly, quite as much so as colour does in Buff.

The desired colour is one uniform, lustrous beetle-green, as seen in the Langshan and Black Hamburgh to the greatest perfection. The under fluff should be black down to the skin, but it is very rare that we can find a bird perfect in this respect. All the points of head, face, wattles, and ear lobes are the same as in all other Cochins, bright red, neat, smooth, and even. The eye in Black Cochins varies more than in any other colour, some being very dark brown—this we think as grave a fault as a white or pearl eye. We think the eye of the Black Pekin should be red, or at least yellow. In breeding Blacks, it has often been noticed, that it is most difficult to obtain the most perfect coloured cockerels and pullets from one pair of birds, the rule being that all the most brilliant coloured pullets' brothers have, more or less, red feathers in their hackles, backs, or saddles; whereas, all the soundest and best Black cockerels' sisters are wanting in lustre or sheen, and look quite inferior in colour to the pullets bred the other way.

Where there is ample room for the purpose, we advise, therefore, that even in starting from one common parentage, two distinct strains should be built up, the one for producing cockerels free from red or straw coloured feathers, and using for this purpose only the deadest black pullets or hens, mated with a sound black cock, and avoiding the more lustrous hens or pullets.

And, on the other hand, we should select the most lustrous beetle-green winged and breasted cock, however much red he showed in neck, back, or wing, and mate with him the most brilliant coloured hens or pullets—provided always that other essential points were sufficiently in evidence.

This has been our own plan, and under it we have succeeded in breeding several cup winning Black Pekins at the Palace, including the three years, 1889, 1890, and 1891, and we are still working on the same lines.

One point we desire to guard our readers against, and that is allowing white ear lobes to gain the upper hand. We remember the time when a judge practically would disqualify (*i.e.*, he would never give a prize to) white-lobed Cochins. We look upon this as a question of degree, and of white lobes, we say, the less the better. Decidedly the lobes ought to be red; but as it is a fact that some of the very best blacks ever bred, including one Palace Cup winner, recently had more or less white in the ear lobes, we should be specially careful never to mate together two birds, cock and hen, showing this fault. We must always try to counterbalance faults in any one point, by extra good points in the corresponding part of the bird mated with the faulty one.

And we will just mention one other point which needs special watching, when mating up one's birds, and that is the feathering at and around the hock joint. Our advice is, whenever practicable, see that this is ample, for it seldom happens that when there is plenty of the right sort of feather here, there is much deficiency on the toes.

Before taking our leave of Black Pekins, it may be of use to some of our readers to know that we have twice crossed between pure Black Cochins and Black Pekins, and by making two crosses in the year, we have succeeded in reducing the birds down to Bantam size, and good enough to win anywhere. With one cock bred in this way, we won the Crystal Palace cup three years ago, and he has been the sire of two cup-winning Blacks and two cup-winning Cuckoos at the same show.

We now pass on to the Cuckoo Pekins. These are a very recent introduction; in fact, the first time that a pair of this variety was exhibited was at the Bawtry Show, in September, 1888, when Master Frank E. Entwisle exhibited three pairs of them, which produced quite a sensation, one pair winning the silver cup. In colour they were quite as perfect as any shown since, though they are now much improved in shape, cushion, softness of tail, and abundance of foot and shank feather. We first produced Cuckoos in this way: While crossing Black Pekins and White Booted with the double intention of strengthening the Blacks, and producing White Pekins, we reared, amongst others, one a rather dirty looking white, so very excellent in shape, etc., that we thought it good enough to show as a White Pekin at the Dairy Show; so we had it caught and washed, but to our surprise it would not come a better white than when first put into the soapsuds; we tried a thorough good soaking, washing, and rinsing, and then had her carefully dried; and on the following morning we had a careful look at her, when we discovered faint but regular bars of stone colour, on a milk-white ground. We at once saw that in this pullet we had a more valuable prize than a pure white would have been, and we mated her with her sire, a Black Pekin cock, for the next season. From this mating we had distinct cuckoo markings, and these pullets we mated with a Cuckoo cockerel, which Mr. Leno kindly sent us, and which he bred from his imported Chinese Cuckoo cock, we believe the only one ever sent from China. Then we bred in-and-in, and back to the pure Black Pekins, until they have proved themselves capable of, now and then, beating all other colours of Pekins.

In the year 1890, our Cuckoos were so vastly improved, that at several shows they scored well, and we were persuaded to part with several pens, among which were the winners of both cockerel and pullet cups at the Crystal Palace, where classes were provided for Cuckoos.

A brother of the above birds won the cup at the Palace Show last season, under the late Mr. Beldon; this bird being considered perfect in every point except size, and he certainly is a trifle large, but we are breeding from him again this year, and expect great things from his produce. Last season we mated some of our best Black Pekin hens with this cock, and among the produce were some of the most perfect shaped pullets we have yet raised; but instead of being either Black or Cuckoo, as all had been previously, there were several that our young people call Magpies, because they were black and white irregularly marked. These are mated with the most perfect shaped Cuckoo cockerel for this season, and the produce will be kept separately from the other Cuckoos to note the result. In our Cuckoos we very rarely have any red or straw-coloured feathers. Should any show themselves in a cockerel, though objectionable in a show bird, it need not, in our opinion (though we know that this opinion is contrary to the generally accepted theory, and that because so many people do not keep separate and distinct pens for Cuckoo cockerel breeding and Cuckoo pullet breeding, and consequently wish to avoid, as far as possible, any straw or red colour in their cockerels' hackles or saddles), be considered any fault in a pullet breeder, rather it may be taken as an indication that the cockerel would be able to increase richness of colour in the bars of his pullets, just as in the case of the Blacks. In breeding Cuckoos with Cuckoos, it may be safe to proceed in this way for four or five years together, but then it is advisable to use pure Black to keep up the proper depth of colour.

Cuckoo Pekins should have very sound orange yellow legs, and orange beaks are generally preferred, though personally we do not dislike a little dark marking on the beak of a Cuckoo Pekin, as it seems quite in harmony with the feathering.

And now we must try and describe the colour and marking of the Cuckoos. These points vary very considerably from a pale, almost white ground, with cloudy and indistinct markings, to a beautiful soft French grey ground, with dark slate bars. The more clearly defined, and the finer the markings, the better. Not only does the ground colour vary, as well as the colour of the bars, or markings of the feathers, but also in different birds the pattern of the markings varies considerably. There are Cuckoos shown with the same pattern of markings as the dark Brahmas and Partridge Cochins—concentric circles of pencilling, one within the other, *i.e.*, in the hens—but this is not correct. The marking we require in Cuckoo Pekins is a series of clearly defined bars (and we prefer narrow ones) across each feather, from the head, down the hackle, breast, thighs, wing, back, saddle and tail, and, in fact, each feather throughout the whole bird, both cock and hen, must have this distinct barring, or series of bands, across the feathers. In some birds, we have counted nine bars across the hackle or saddle feather of a cockerel, but seven bars make the feather look well. A less number would not be so good. In hens, across the saddle feathers, five bars are sufficient, and as the feathers on other parts of the body are not so long, a proportionately less number of bars are required. The same description of marking is required on the feathering of the legs, feet, and toes, and the more distinct the better.

A common failing of Cuckoos is to have some of the wing feathers white, or with a good deal of white in them, and also in the tail feathers. This is a grave fault, and is reproduced in the chickens most persistently. If the fault is seen in the brood cock, it will not do to run hens with him having the same fault. If the cock bird is perfectly sound in colour, less anxiety need be felt about a little white in the hen's flight feathers. Never breed from any—either cocks or hens—that are broad, coarse, or irregular in their markings.

Partridge Cochin Bantams are quite a recent production, and, as mentioned previously, our first was the result of a cross between a Black and a Buff. This was, however, much too dark in his hackle and saddle, and the feathers were almost plain red, instead of gold, striped with black. For the last four years, we have bred cockerels as perfect in colour as any of the large breed of Cochins, and these are the result of breeding down from small specimens of the large Cochins, which we obtained from Lady Gwydyr, Mrs. Goodall, Mrs. Turner (of Bath), and one or two others.

But, in order to make our paper on this subject as practically useful as possible, we will first try and describe the ideal exhibition colour and markings of the cock and hen, and then show how to produce such birds from breeding pens, in order to do which, we must have two separate and distinct breeding pens.

The Partridge Cochin Bantam cock must be the exact miniature of the large Partridge Cochin—*i.e.*, in colour. His breast, tail, thighs and underparts must be a sound black, and the more of the beetle-green shading on the surface of the feather, the better; the sickles, tail coverts and wing-bars, bright beetle-green; the neck and saddle hackles bright orange, inclining towards golden rather than towards red, each feather distinctly striped with black; the head feather, and upper part of hackle, a shade deeper in colour than the lower part, which falls over the shoulders and back; the shoulder butts, black; the wing-bow and across back, including the shoulder coverts, a rich, full crimson, and this should almost cover the wings, being broader and larger than the corresponding crimson portion of a Game cock's wing, owing to the feathers being so much softer, longer and broader; the flight coverts, a good full chestnut or bay, as in the Black-red Game. The foot and shank feathers should be solid good black, free from either rustiness, or grey or white feathers; of course, the legs must be yellow, and the toes yellow. The colour of the hen or pullet must be a clear light golden-brown ground colour all over the breast and body, legs, feet and wings, with every feather evenly laced with concentric rings—one inside the other—of very fine pencilling of black, or sometimes very dark brown. We think greenish-black pencilling on a golden-brown ground looks best, and that is what we aim for. The hackle and head are plain golden, or straw coloured, clearly and evenly striped with black. With this character of colour and markings, other points being equal, anyone need fear no rival.

But it would not be likely that anyone purchasing a pair such as the above, and mating them together, would find either cockerels or pullets bred from them come quite up to the perfection of the parents, unless in a few cases; and if the produce were bred in-and-in on the same lines—that is, mating the most perfect-coloured cockerels with the pullets—there would be less and less distinctness in each succeeding generation, in the pencilling of the pullets bred.

We know breeders of skill and care, who have set themselves the task of making a strain that would breed both cockerels and pullets true to the requirements for successful exhibition; but after years of hard and patient work in this direction, have given it up as not worth while proceeding further with; and, in fact, have acknowledged that "it can't be done."

How, then, can such perfect coloured birds be bred, as some of the large Cochins we saw shown last season? To begin with the cockerels. It is necessary that the cock or cockerel heading the cockerel breeding set of hens be as perfect in colour as possible, sound black in the breast, right up to the throat and down to the toes (a very common failing being a patch of brown feathers in the centre or upper portion of the breast, and a few grizzled or rusty markings on the foot feather and fluff). These faults should be avoided in the cock from which you want to breed perfect-coloured cockerels. Next, your hens must be chosen from a known cockerel breeding strain, and the freer from the good bold pencilling required in the show pullet, the better. We have known the most perfect coloured cockerels bred year after year, and birds which moulted sound black in the breast the second and third years, from hens that were just simply the finest marked dark grouse-coloured all over body and breast, but a good, clear, bright colour in the hackle; and we have known equally good cockerels bred from hens with fairly marked wings, but lacking distinctness of pencilling on thighs and feet, and backs, etc. Our choice is always for the more sober coloured bodies, with distinctly striped hackles; and when we have these points, we give all the rest of the consideration to points other than colour, on which we are now writing.

And now we will consider the breeding pen for pullets and hens. Here we want a different sort of a cock bird altogether. Just in two respects he must exactly resemble the male in the cockerel breeding pen—that is, he must be perfect in shape, and in quantity of feather. In colour, we can best describe him as a very bad coloured bird—he must be very rusty in breast, thighs, shanks, feet, and all other feather underneath, and dull in his flight coverts, and more heavily striped in his hackles; the top colour as sound as you like, but he is all the better if his wing-bar is laced. With such a cock, provided he shows good breeding in other points, such as quality of head points, shape, size, abundance of feather, soft tail, etc., you may safely put the very best show hens you can meet with, and rest satisfied you have a good pullet breeder at the head of affairs. Now, supposing we were to start with one perfect pen, as regards colour and markings—let us say a pair of cup winners at some good show—it would be easy enough to make, from this one pair, two distinct strains which would breed true to points required—*i.e.*, provided the cup-winning pair had been produced by careful breeding on the generally adopted lines, and were what we should understand by the term, high-class birds.

And we will explain this, in order to prove that it is quite a mistake to say that "it proves bad breeding when two distinct strains of a breed are required to produce one prize pair of opposite sexes." Let us say, for example, that we start with only this pair of cup-winning Partridge Cochins, and the hen lays a dozen eggs, and then comes broody, and is allowed to sit, and hatch and rear the chicks; and let us suppose that she rears ten—six cockerels and four pullets—we should then proceed as follows, so far as breeding for colour goes (and in this example that is the only point we are trying to illustrate):—When the cockerels were about twelve or fourteen weeks old, we should carefully examine the lot, and select the two most nearly like the colour and marking of a pullet, *i.e.*, pencilled on breast, thighs, fluff, foot feather, shoulders, wing-bar, and flights; then mark them for final selection as pullet breeders, killing off the other cockerels as no good; and when the above two were in full feather, say twenty-four weeks old, then finally choose the more rusty coloured of the two, *i.e.*, rusty in breast, thighs, feet, etc.; and for the second year mate him with his mother only, and proceed on the same lines, at the season's end selecting the most rusty cockerel which showed best pencilling in his young days. This bird might breed with both his mother and sisters the following year, and establish a reliable strain of pullet breeders, always selecting the best pencilling in both sexes for the reproduction of pencilling.

Now let us for a moment retrace our steps to the first four pullets bred from the cup pair, and from these we wish to establish a cock breeding strain.

Our only care in selecting from these four would be to choose the two best in shape and size, and abundance of feather, and we should entirely disregard pencilling, preferring to depend on the result of selection for colour in the third and fourth generations, rather than the first and second. Having then chosen a couple of the cup cock's daughters, we should run them with him for the next breeding season, and at six or eight months old, select from their pullets the best shaped and most Cochiny birds to breed back to their sire the following year, for a cockerel breeding pen, and from these we should expect but few rusty cockerels. The influence of the sire —a well-bred cock for generations, always insisting on black breasts—would be sure to tell heavily on his sons.

After this, we should give the preference to pullets, weak in pencilling, for future cock breeders, always being strictly particular to have perfect colour in the stock cocks.

Of course, size is a very important point in any Bantams, and in the case of Bantams bred down from large breeds, cases of reversion are very frequent, but there need be no difficulty on the score of size, if the eggs are set in August or September, and the birds are reared in the winter. We have plenty of September, and some October chickens laying now (April), and some are sitting.

In breeding down from the large Partridge Cochins, both Black and Buff Pekins have had to be used; but in a very short time, we believe our strains will be as fully reliable as are the big breeds, for the reproduction of prize winners.

White Cochin Bantams are just what their name implies. They are now just so far perfected, that any one can breed winners out of nearly every brood of chickens. The chief difficulty is to keep the birds out of the sunshine, which spoils the colour, so quickly turning it yellow. Where the birds have a good grass run in the shade, and are not exposed to the damp, they will maintain the purity of the white as well as any white breed will, or can. If ever we require a cross for our Whites again, it will either have to be with a pure white Cochin cockerel of the large breed, or else a Black, or a Cuckoo Pekin hen. We crossed once with a splendid shaped White Cochin cockerel, kindly sent us by the Rev. C. D. Farrar, and all our present White Pekins are descended from that cockerel, which afterwards won first prize at Paris Show some three years ago.

We prefer those Whites which, when new hatched chickens, are a sooty shade (as if they had been rolling in a soot heap, and consequently looked quite greyish), rather than those which are of a yellowish tinge or shade. We find these latter are more liable to grow up of a yellowish white.

Our impression grows stronger that White Pekin Bantams are destined to become as general favourites as any white breed of Bantams, and that before long. Though it has always been considered a difficult task to breed pure white Cochin cockerels, with good orange legs, it has been done; and what has been done can be done again, both in large and small White Cochins.

BRAHMA BANTAMS.

We must now leave the Pekins, and turn our attention to what we have heard called their cousins—the Brahmas.

Brahma Bantams are, like the large breeds of the same name, of two varieties—Light and Dark. It is something like seven years since we first ventured to exhibit our Light Brahmas. Previous to that time, we always thought them too large to call Bantams, but as we were successful in winning a few prizes, one or two other breeders took them up, the most successful of them being Mr. Astley, who has won a Crystal Palace cup with a very nice pullet; in the year 1889, we think it was.

Light Brahma Bantams are still in few hands, and need careful breeding before they are likely to become as popular as Pekins, and yet they are striking birds. What can be prettier than a flock of half-a-dozen snowy-white bodied pullets, with their black beaded necklaces, and neat coral-red combs, and white heads, with neat, little black tails just edged with white, and led by a good coloured Light Brahma cock, with his red, coral-like comb, and snowy head and breast feathers, well-tipped neck hackle, white body and dark tail? There is something very striking in the appearance of a good flock of Light Brahmas strutting across a lawn, or a well-kept paddock. Then, too, they are more hardy than most of the Bantam tribe; lay larger eggs, in proportion to their size, than most do, and also lay a large number of them. As we were among the number of those who bred the large Light Brahmas seventeen or eighteen years ago—our birds being chiefly reared from the strains then in the hands of Lady Gwydyr and Professor James Long, from the latter of whom we had several sittings of eggs, chiefly from sisters of a noted cock bird weighing over 18 lbs.—we had a few years' experience in seeing how both Dark and Light Brahmas could easily be produced from one pair of birds. That, and other experiences we had with them, have not been forgotten, but turned to good account in the evolution of Brahma Bantams.

It may be useful here to record the process by which we first obtained Light Brahma Bantams. We had, from a gentleman in Devonshire, a very small, but typical, Grey Aseel cock, which for a time ran in a yard with small Cochin hens—White, Buff, Partridge and Black—and a few White Booted Bantam hens. From these we raised a large number of chickens, intending to use the pullets for sitters and mothers, and to kill the cockerels for table use, knowing that this cross ought to produce the very best mothers of all. From this Grey Aseel cock, and a White Cochin hen, we reared some very pretty pullets that would have been pronounced fair Light Brahmas by any judge. All had the Brahma or Aseel comb, and though rather dark in fluff, and marked on back, were not more so than many so-called pure bred Light Brahmas. From the Black Cochin hen we had several dark greys, some marked like Dark Brahmas, but more of them like the heavily-marked Birchen-greys, and nearly all these had the true Brahma or triple comb, and were not far off the correct shape of Brahmas. What surprised us most, was the length of the hock feather, and the quantity of shank and toe feather which these chickens possessed; but, of course, they were too leggy, and too narrow at the stern for Cochins, though not bad Brahma shape and style. The Dark and the Light Brahma colours so produced, we have bred together, and have proved that it is comparatively easy to produce the beautiful steel-grey colour now so much admired in the Darks.

Finding how nearly we had got Light Brahmas, we mated the same Grey Aseel cock with some yellow-legged White Booted Bantams, and in the first cross there were several chickens the colour of Light Brahmas, but also others of all colours, from buff, partridge, dark grouse, to black. In the next cross—*i.e.*, between the cockerel of the Aseel White Cochin cross, and the Aseel White Booted cross pullets, all of good Light Brahma colour, and from which we confidently looked for the retention of colour and markings, with an improvement in shape—we found nothing but disappointment, the produce nearly all reverting to their very ancient ancestors, and being chiefly brownish birds, all the cockerels of which we killed.

About this time, we had a small but typical Light Brahma cockerel from Lady Gwydyr's yard, and mated several of these mongrel-coloured pullets to him, producing the right coloured Light Brahmas in their chickens. One point that the Aseel was very useful in producing, and perpetuating, is the neat and small pea comb, and another point is the bright yellow leg. Scarcely any variation in these two particulars came under our notice. There were, however, two points which needed alteration—first, the tails were too long, and secondly, the size was too large in most of the birds thus bred. So to reduce size we resorted to in-breeding, both early and late in the year, generally preferring to use the smallest cockerel in our lot.

We also obtained from O. E. Cresswell, Esq., one or two Dark-tailed Light Japanese Bantam hens, and reared a lot of chickens from them, and one of our Light Brahma cockerels, with the object of reducing size; and this we certainly did, rearing quite a lot of chickens of proper Bantam size, and good sound colour, with nicely feathered yellow legs; but, oh! what tails and what combs! Well, we thought, we can't have everything at once, so we made up a second breeding pen of the best of these Brahma-Japanese cross pullets, with the most typical Brahma cockerel, son of Lady Gwydyr's bird, and we bred from them on a walk away from home, and bred their pullets again, next season, back to their sire. By this time, they were really like good Light Brahmas, and won prizes as such. On the other hand, we persevered in breeding the others without any of the Japanese cross in, and we must say that we so greatly prefer them, that our present stock does not contain a single bird of the cross containing any of the Japanese blood, our greatest objection to which is the great length of their tails; and every now and then this will re-assert itself in the offspring with great pertinacity. These were, however, often successful in the show pen, and several fanciers pressing to have them, we parted with a few pens, and, among others, some went in the year 1886 to R. B. Astley, Esq., who bred some beautiful chickens from them, beating us in the show pen a time or two, both in colour and size; others went to America, sailing in the *S.S. Indiana* (Capt. Boggs) from Liverpool, November, 1887. Others went to Holland, Belgium, and France, and a few among home breeders. This season we only mated one pen of them, and are rearing but a few broods. But we are glad to number among recruits, Mr. Rowland Butterworth, in Cheshire, and Mr. J. W. Edge, of Birmingham, so we hope there will be some competition among them in the Autumn, especially as there are hopes of classes being provided specially for Brahma Bantams at the large shows in the Autumn. Brahma Bantams ought to be, of course, the exact counterpart of the large Brahma fowls, only in miniature; size should be the only difference, and the Bantams should be about one-fifth the weight of the big ones. We will, therefore, now proceed to give a description of them, and next describe the theory of mating to produce the required points.

The Light Brahma Bantam cock: Head, pure silvery white; hackle, the same shade, but in the lower parts distinctly striped with sound black (a very common failing is grey); the breast, belly, thighs, shoulders, wings, and back, etc., white; if possible, white down to the skin, and this the Americans prefer; but frequently, if not generally, the fluff or down is dark. This is not objected to, if it does not show through the feather. The saddle looks best when slightly, but evenly striped like the hackle. The leg feathering as white as possible on the surface, and the same with the toes, but there should be distinct black in the underparts of the feather. The wings, when open, should show black in the primaries, as well as black on the inside of the secondaries. The tail should be black, the top outer feathers and the sickles slightly edged with white. The skin of the shanks and toes, and the scales should, like the beak, be bright orange yellow. The hen should at first sight appear white all over, with exception of a dark-beaded hackle and dark tail. The head, white; neck hackle, white, distinctly striped with black; body, breast, thighs and wings all white, except that the wing primaries are black, and the under side of the secondaries also black; tail, black, with white edges to upper feathers; shanks and feet well covered with feather, chiefly white, but what black is visible should be distinct. In shape they should not resemble too nearly the Cochin or Pekin, being longer in leg and neck, higher in cushion and tail especially, and more erect and commanding in general carriage. The eyes may be either yellow or red; we prefer red, as indicating more vigour than a pale eye.

And now as to the mating of the birds. It will be found easiest, and always most reliable, if two strains are bred, one for cockerels, and the other for pullets, and having previously explained why, we make no excuse for at once indicating how we should proceed to mate the birds. First, to breed cockerels, select the most perfect coloured cock or cockerel you can obtain, very distinct in hackle, and nicely striped saddle, and with plenty of feather, and good shape, and put with him hens as white in wing, body, and back as possible, even if rather wanting in colour of hackle. To breed pullets, we should select as light coloured a cockerel, clear in body colour, legs, and saddle as possible, and only slightly striped in hackle, with the darkest hackled and blackest tailed hens possible to obtain, with white backs and wings. Though it is not certain that we shall in this way get all we want in colour, there is no more reliable rule that can be followed, and we had better follow it, and get the best results we can in this way.

Dark Brahma Bantams should be simply, on a greatly reduced scale, the exact facsimile of the exhibition Dark Brahma fowls, but before they are brought to such perfection as this, both time, and many careful times of in-breeding, may have to be resorted to. Up to the present time, the smallest adult Dark Brahma cock we have bred and won with, was the winner of first prize at Bournemouth, in August, 1891, which, with his daughter, twenty-two weeks old, won first in a variety class. The pair weighed about 3 lbs., and in colour, the pullet equalled many winning large Brahmas. The cock was of a rather yellow cast in hackle, and not quite so clear white on back and shoulders as some of his sons, which are equal in colour to any large Brahmas seen in the show pen. These Bournemouth winners, along with another pullet or two, were sold to the Count of Chabaunes, in France, who won a first prize with one of the pullets at the late Paris Show. We have won other prizes with pullets at the Dairy Show, the Palace, the Bantam Club Show at Kendal, and Portsmouth, the weights of these pullets being 20 to 23 ounces, at $5\frac{1}{2}$ to 6 months old, when they had just acquired their new feathers.

It is so far satisfactory to be able to state that the pullets are now produced with a very fair amount of regularity in their markings, and perfection in colour, and that there is reasonable prospect of the cockerels breeding quite as well this season. We have three pens mated for breeding chicks of this variety, and a most beautiful lot of chickens growing up. As, however, the entire stock of Dark Brahma Bantams at the present time in existence is so very small, and confined to so few hands, we can but select from such birds as are available; but it will not be very rash to predict, that ere ten years have passed, they may be counted by hundreds, where now they scarcely number their tens. We will next describe the standard colours, and then the mode of procedure in selecting and mating the breeding stock.

The Dark Brahma Bantam cock should have the feather of the head silvery white; neck hackle, white, clearly and distinctly striped with black, all white at juncture with the head, but the stripes increasing in breadth down to the shoulders and back. We prefer a well-striped (*i.e.*, rather heavily striped) hackle; the back and wing-bow, pure silvery white; saddle hackle, white, well striped with black, and the black stripes increasing in breadth on to the tail coverts; the breast, belly, thighs, shanks, and toe feathers, as black as possible in a show bird, or a cockerel breeder (but the pullet breeding cockerels are all the better for a white ticking on the breast, thighs, and shanks); tail, black, but if the sickle feathers have a narrow, distinct white lacing round the edges, it is admired, and considered a point in their favour by many breeders and judges; the shoulders and wing butts are black; wing-bars, rich beetle-green; side sickles and tail coverts, also beetle-green; shanks, orange-yellow; beak, dark horn; wattles, lobes, and comb, coral red; the comb should be smallish, and triple, like a good Aseel's comb; eyes, the redder the better.

The exhibition hen should be white on the crown of the head, and very clearly and evenly striped with rich black on a silvery white ground in all her hackle feathers: a good, full, broad stripe of black is to be preferred, and the marking to come well up to the throat, and level with the under side of the eye, at back of neck; any weakness here is almost certain to be accompanied with deficiency in pencilling on upper part of breast; we should, therefore, lay great stress upon a good hackle. The tail should be black, with just the outer edges of the top feathers slightly marked with grey; all the rest of the body should be of one uniform shade of colour—a light French-grey ground, with steel-grey pencilling, looks as well as any shade. Some breeders prefer a darker shade of ground, almost a light slate, with black pencilling, and this is a colour which lasts better, and does not fade or look sunburnt so soon. Formerly a light drab, and sometimes brownish shade of ground colour, with black pencilling, was successful; but whichever shade of colour be adopted, uniformity of colour all over must be bred for, and insisted upon, in the pullets, and the more even and distinct the pencilling on the breast, right up to the throat, and down the thighs on to the shanks and feet, the better; the same marking also extending over the wings, back, and saddle.

It will be found that, as pullets moult, at the end of the first season, in ninety-nine cases out of every hundred, the new feathers will be of a larger and bolder pattern of pencilling than that of the first season; and in very many cases the shade of colour in the second moult will be of a more drab or brown cast than in the first year. We have a decided preference for a steel grey or bluish grey

BRAHMA BANTAMS.

DARK COCK.

Bred by, and the Property of, Mr. J. F. Entwisle.

Winner of First and Cup at Crystal Palace, and Third at Birmingham, 1893.

DARK HEN.

Bred by, and the Property of, Mr. R. Butterworth, Pownall, Wilmslow.

"Pownall Pride," winner of First Prizes at Crystal Palace, Altrincham, Reddish, and First and Cup at Fairfield, 1893.

PAIR OF LIGHT BRAHMAS.

Bred by, and the Property of, Mr. J. F. Entwisle, Calder Grove House, near Wakefield.

Cock, winner of First Prize at Glossop; Hen, Second at Crystal Palace, Leominster, and Third at Birmingham, etc., 1893.

shade all through, with rich black lacing and pencilling. The pencilling itself is a point that varies very considerably, from a very fine mossy marking, through varying ranks of distinctness, up to very broad black bands or concentric rings one within the other, the entire length of the feather. What is termed the happy medium is the choice of the majority of breeders in the large Brahmas, and, we think, should be in the Bantams also. It has been our aim to produce this type, and, having obtained it, we mean to keep to the one type and try for no other; thus fixing the character of the breed. We have now lying on the desk before us, feathers taken to-day from the breast, wing, and back of one of our prize hens, now about 2¼ lbs. weight, and too large to win this season; but, as a pullet, one of the most beautiful we ever bred; and in these feathers, each one has five distinct rows of pencilling, nearly black, on a light French-grey ground, so that there are, counting the outer edge (which is the light shade), six rows of this shade on each feather, and in this bird there is the same distinct pencilling even on the fluff, and under the tail, and on the shanks and toes. Among her chickens we think we can find half a dozen of the same character, her own son being her mate this season. He is slightly ticked on breast, and on shank, toe, and thigh feather.

Now let us just consider the mating of our breeding pens, in order to maintain these points of colour and markings, and it will be found that, in order to be successful, the same principles or rules of procedure must govern our choice, which we observed in mating the Partridge Cochins; and, as we explained at considerable length the why and the wherefore of this matter, we must refer our readers to what we had to say only a few pages back on Partridge Cochin mating, and breeding. It will, however, be found a general rule, that to insist on perfectly black breasts, thighs, shanks, and foot feather in either Partridge Cochins or Dark Brahmas, we are choosing birds calculated to breed less pencilling on the backs, breasts, and underparts of the pullets. This much may be taken for granted, and the exceptions to the rule will be but few. In choosing a pen, therefore, for producing Dark Brahma Bantam Cockerels, look for perfection in every point possible in the cock bird that is to head the pen; and with him mate the smallest, best shaped, neatest headed, best combed (and be very particular on head points) hens or pullets you can procure, with perfectly silvery heads (no brown caps on any account), very silvery hackles, and plainish, *i.e.*, lightly pencilled breasts and backs, and on no account any white showing in foot feathers.

With these you cannot get far out of the direct road to success, especially if you have plenty of feather, and good shape in each sex.

But to breed the best pullets, as we have said before, you must look for the highest type of perfection in your brood hens; and if they are everything that they should be, the best cockerel or cock to mate with them will be a son or a brother of one of them, *i.e.*, one bred from a perfect show hen; and we should look for the following points in him: smallness of size, combined with cobbyness, neat comb, sound red lobes, and red eyes, purity of silver, and distinctness of stripes in hackles (a cloudy or mossy hackle we detest), evenly ticked breast, thighs, and foot feathers and fluff; very silvery back, saddle and wing-bow; well striped saddle hackles, and as much shank and foot feather as we could possibly get, without having ugly vulture hocks; and from such a pen we should confidently look for the very best possible results. This breed is well worth taking underhand, and there is at the present time an opening for any skilful breeder with time and patience, and room at his command, and without the expenditure of much cash, to build up, and establish, a strain of what could not possibly fail to please good taste in colour, shape, and feather.

BOOTED BANTAMS.

Booted Bantams are of very ancient origin undoubtedly, and most probably have existed for many centuries. Our recollections of them carry us as far back as 1841, when we had a present of a pretty little pair of Black Booteds. They were not only our first fowls, but our first live pets of any kind, and our trouble was great when, several months after, we one day found the little cock bird dead in his house. The hen lived for a long time, but we never got her a mate. Booted Bantams, now, however, are like other breeds, much improved, and there are more varieties of them now to be had than even a few years ago. The following are tolerably well-known varieties:

Black, White, White-whiskered, Speckled Black and White, Speckled Red, Dutch-bearded Booted; closely following on which are the Burmese, Sultan, Silkies, etc., which shall be fully described in due time.

For the present, however, we must take the best known varieties, viz., White Booted and Black Booted. Both these varieties are considerably larger than the general run of Bantams, having been bred more for feather than for size, or, as some affirm (and we should not care to say they are altogether wrong), they have been allowed to breed anyhow, with no definite set purpose in the mind of the breeder inducing him to model the breed to suit his own taste. The result is that we meet with them of totally different types, some closely approaching the Japanese in shape, and also in carriage of wings and tail, but long in shank, so as to carry their long vulture hocks just off the ground; others short in leg, and with higher carriage of wing, and lower of tail. This class finds most favour with judges and breeders alike at present time, and we think will continue to do so. This, too, is more of the character of the White-whiskered Booteds so successfully shown by Mrs. Ricketts, of Knighton, and frequently beating the other Booted in open competitions. The best Blacks, however, of late years, have been of a larger and more upright type, and very full and heavy in tails. Mrs. Ricketts and Mr. E. Walton have shown some of the best of these Blacks.

White Booted Bantams, as we stated before, are a very old breed, and formerly were as frequently shown with yellow legs and bills as white-legged. There is no doubt, however, that now white legs and beaks are considered the proper colour, and if the two varieties were competing side by side, and of equal merit in every other point, no judge would for a moment hesitate in placing the white-legged birds above the others.

The comb must be single, well serrated, and perfectly erect in both cock and hen; the ear lobes, red (any white being a fault), neat and small lobes to be preferred; wattles, small and close, not long and Cochiny; shanks, heavily feathered, with long and rather stiff feathers; toes as heavily feathered as possible, and the longer and more closely feathered the "boots," the better; the feathers above the hocks must also be long and full, and if the tips of the hock feathers touch the ground as the bird walks, so much the better. The wings are long and generally carried drooping, which is natural to the breed, though there are fanciers who desire to see high carriage of wings. We, however, think it better that there should be no attempt to breed White Booted Bantams away from the old type, and so nearly approaching the White Pekin type, as to puzzle anyone to decide which to call them. The White Pekin has short wings, carried high, and clipping the body tightly, making the fluffy cushion stand out all round and over them. The White Booted hens' wings are longer than the body, and their tips pointing toward the ground. White Booteds are long in the tail, and the cocks should have "full tail furnishing," *i.e.*, plenty of sickle feathers on each side of the tail. The colour must be pure silvery white, and to keep the colour pure, the birds need plenty of shelter from sun, wind, and rain. They also require a very smooth, level floor to their house and run, to prevent the foot feather being broken. In weight, the cocks are generally 20 oz. to 26 oz., and hens 3 oz. or 4 oz. lighter, but 16 oz. is quite little enough for any hen to weigh when in show form.

Next, as to breeding. We have found no advantage in keeping separate pens for breeding the different sexes, and as the standard is, it will be found that both winning cockerels and pullets can be bred from the same pair of winners, therefore no particular direction need be given on this head; but we may just say that April and May will be found the best months for hatching the chickens, and the best food will be such as is advised for Bantams generally. A White Booted hen will generally sit, and cover about seven of her own eggs, and we have found them very good mothers, and seldom break or roll their eggs out of the nest, as one might easily expect them to do with such heavy foot feathering.

White-whiskered Booted are in all respects like the ordinary White Booted, excepting that their cheeks are provided with abundant feathering (and also under the beak, like a beard), this feather being about ½ in. to ¾ in. long, and standing out like whiskers, thus giving the distinguishing name to the variety. These are very rare, and we do not remember seeing them exhibited except by Mrs. Ricketts, and two or three others. We found them most difficult to rear, and many of the eggs

SPANGLED BOOTED BANTAMS.

Bred by, and the Property of, Mrs. Ricketts, Knighton Vicarage, Radnorshire.

Cock, winner of First Prizes at Birmingham, Liverpool, Portsmouth, and Second at Crystal Palace, etc., etc.
The Hen also a Prize winner.

unfertile, no doubt owing to persistent in-breeding on account of their scarcity. It would be a pity to lose altogether such a quaint variety.

No doubt their constitutions would be invigorated by the introduction of fresh blood, but that could only be done at the expense of some other points.

Our plan would be to use the best White-whiskered Booted cock with ordinary White Booted hens, and save the pullets so bred to breed the following season to the pure Whiskered cock, and the Whiskered hens to the son of the Whiskered cock and plain Booted hen, and then breed back to the pure Whiskered strain again.

Of course, by this cross, we must expect to diminish the quantity of whisker, and probably lose the beard altogether. Both points can be recovered by resorting to the Sultan for a cross, but probably most breeders will be content to breed them more pure than that, and content themselves by the best results they can procure from careful selection of the produce of the first crosses. We may add that we have never seen the White-whiskered Booted with any but the whitest legs.

Black Booted are handsome Bantams, rich and lustrous in colour, black legged, with black beaks, red combs, lobes, wattles, and faces, and generally they weigh an ounce or two more than the Whites do, and they look larger.

The points are just the same as in the White Booted, except colour, and in this, great care must be taken in order to keep up the purity and richness of colour, as seen at the present time.

In order to do this, it will be at least desirable, if not quite necessary, to keep two breeding pens, and follow exactly the same directions given for breeding Black Pekins, to which we refer our readers, rather than repeat the directions again. The main point to avoid in the show bird is colour in the cock's hackle, back, or saddle, and this can best be done by breeding from the duller black hens and a sound black cock, to breed cockerels; and, on the other hand, the bright beetle-green shade in the show hens requires as much attention to keep it up, and the best pullet breeding cocks often, if not generally, have a little red or straw colour in them.

Black Booted Bantams often have very dark brown eyes. They should be either red or brown —both colours are admissible. They should be very heavily feathered in tail, and on legs and feet; and if the birds can be shown under 24 oz. for old cocks, and 20 oz. for cockerels; 20 oz. for hens, and 17 oz. for pullets, they will be small enough to win.

The Speckled, Splashed, or Spangled Booteds are merely variations in colour, and are not general favourites; though, as tastes differ, there will be fanciers who will give the preference to the Spangles.

In colour, the Black Spangles are just like Houdans, while the Reds resemble very closely the old Speckled Dorkings, or Black-reds that have been out in a snowstorm, until half covered with snowflakes. There are also tri-coloured Booted, which, like the Bearded, are chiefly of Dutch or Belgian origin. Many of them are marked very much like the Almond Tumbler pigeon.

SULTAN BANTAMS.

Next on the list are the Sultan Bantams, which we produced by mating together some of our White Polish Bantam hens, and some White Booted Bantam hens, with a small Sultan cock. We placed both kind of hens with him in our orchard at Newington House, so that we might have the double cross—*i.e.*, the Sultan blood on each side in the two families we wished to breed together the second season, well knowing that the Sultan-Booteds would be poor in crest; and the Sultan-Polish good crested, but poorly booted; and then we had to breed in, and only once or twice breed back to the pure Sultan blood.

To our surprise, most of the first cross chickens had four toes instead of five (natural to the Sultans). In this case, the usual rule of the progeny resembling their sire in outward character was varied; and as we did not rear a single cockerel from either side which quite came up to our expectations, we proceeded as follows, breeding from two sets or pens, viz.: First, we placed only two pullets, one a Sultan-Booted, and the other a Sultan-Polish, with their sire, being undecided as to which would produce the more valuable cockerels for future use; but we were determined to breed from a cockerel three-quarters pure Sultan, with pullets half Sultan, quarter Polish, and quarter Booted. In the other pen, we mated a Sultan-Polish cockerel, the largest crested we had, but four-toed, with Sultan-Booted pullets, some four-toed, and one or two five-toed ; and we saved only the pullets bred from this pen, killing every cockerel reared excepting one, which we gave away to breed with some half Burmese Bantam hens. The next cross we made was by mating a three-quarter Sultan and quarter Booted cockerel with some of the smallest pullets bred from the other pen ; and the other smallest pullets with a three-quarter Sultan and quarter Polish Bantam cockerel ; and from the produce of this mating, we reared quite a number of small birds, which we bred in-and-in for some years, and only once or twice had recourse to a pure bred Sultan cockerel for a cross to obtain a cockerel (and never pullets) to continue the breeding. In about six years, by making two matings (spring and autumn) a year, we had a few Sultan Bantams which we could feel proud of, and exhibited at the leading shows. Our object being attained, and having succeeded in producing Sultan Bantams ourselves, we then sold the best pen for a mere trifle, and then cleared out the others among fanciers willing to try and keep up the breed. There are, no doubt, some in various parts of the kingdom, but we have seen less than a dozen during the last twelve months.

BURMESE BANTAMS.

Burmese Bantams we have had in various colours, *nearly pure bred*. The first Burmese we saw were at Bingley Hall, Birmingham, about seven and a half years ago, and after that we were fortunate enough to obtain the cock bird alone (the hens having died) for breeding from ; and mated him with four of our best Sultan Bantam pullets, all pure white birds; but the result were Blacks, Browns, Greys, Speckled and Whites, the latter being in a minority. The pure White Burmese had been sent direct from Burmah by an officer in the British army, to another brother officer in Scotland; but the damp Scotch climate did not suit them, and the last surviver was sent to Newington to save his life.

He was a very quaint little fellow, pure white, heavily crested, with a straight, single, small comb at front of his crest, long winged, very long sickled, his tail being one of the most striking features about him. His legs were so extremely short that his breast and body touched the ground as he shuffled along ; he had heavily feathered legs, and his toes had feathers five inches long, which made his feet look like wings. His weight was very little over 20 ounces. He lived to present us with quite a large family of chickens from his own pullets, and among them some nearly as good as himself.

We also bred some of these crossed Burmese pullets among our Sultans, to the evident improvement of the latter in certain points, but to the loss of the pure white legs ; for *all* the Burmese have orange yellow legs, excepting the Blacks, and their legs were dark willow or black, with yellow under the feet and between the toes. The Burmese are also a four-toed breed. These, too, were scattered abroad, several going to Mrs. Ricketts, others to Miss Arnold, etc., etc.

SILKY BANTAMS.

Silky Bantams are very scarce, much more so than the ordinary Silkies. Real Silky Bantams we have had of the following kinds :—Red-faced Whites and Mulberry-faced Whites, both single combed, and with the ordinary walnut comb ; Golds, Browns and Blacks ; and some of these have not exceeded 22 ounces when full grown. They are most excellent sitters and mothers, but are very subject to scaly-leg disease, and may probably then transmit the disease to the young ones they bring up.

But this chapter has already exceeded its contemplated length, and we must at once proceed to a description of those splendid productions of the late Sir John Sebright, which still bear their founder's name.

CHAPTER VIII.

SEBRIGHT BANTAMS.

THESE are of three colours, Gold, Silver, and intermediate between the two—Creamy. Of these three, the Silvers are the most fashionable, and the most plentiful, Golds coming next, and the Creamies last of all. With the breeding of Gold Sebrights, one difficulty is experienced, which does not apply so much to the Silvers, and that is the ground colour, and this varies considerably in different strains, and in different birds, not only of the same strain, but also of the same brood. But if this difficulty were all, it need not trouble the breeder much; but such is not the case, for, when the time arrives for the chickens to change their first feather, which is generally more black than coloured, and with only a gold or silver shaft down its centre, and if the weather is cold, or damp, or very windy, and the chicken gets the least chill, or drawback in any way for a day or two, it will leave a mark on the feathers of many a Gold Sebright.

If the chickens do not go steadily on with the changing of their feathers, some will probably vary from the others in shade of ground colour, and become what we term uneven, spoiling their prospects of success in the show pen. The principal thing to observe, therefore, in unsettled or unfavourable weather, is that the chicks are well sheltered, and well fed.

The origin of the Sebright Bantams is very well known, and not disputed, but it will not be out of place again to record the fact, that to the late Sir John Sebright, Bart., we are indebted for the production of this beautiful and unique variety. It took Sir John many years to manipulate all the crosses, and so to blend the products, as to finally evolve the most charming of all the Bantam tribes; but with the true Excelsior spirit, he progressed with the breed until he had surmounted all difficulties, and left behind him a new race, which will be a lasting memorial of his taste and skill as a breeder. The birds used as crosses were chiefly Gold and Silver Polands, and Black and White Rosecombs, and Nankin Bantams, though it is just possible that some other birds were also used. But whether or not Sir John did use any other variety we cannot positively say, but this we can affirm, viz., that at the present day, the breed could again be produced (if entirely lost by accident or deliberate intent), by crossing between the above named breeds only.

It is now close upon, though not quite, a hundred years since Sir John Sebright commenced to breed the laced Bantams, and after ten or a dozen years, he, in conjunction with a few of his friends, established a club, which they named the Sebright Bantam Club, the object of which was to improve the breed by obtaining distinct lacing of rich black, on a clear ground colour, either silver or gold.

This club met once a year, on the first Tuesday in February, to hold a show, each member bringing the best birds he had bred in the previous season. The show was confined to birds under one year old, and bred by the exhibitor, no borrowing or lending being allowed. The prizes were contributed out of members' subscriptions, which were two guineas per annum for each variety, Gold and Silver, the Creamies not being recognised as show birds. The standard requirements were, cocks not to exceed 22 oz., and hens 18 oz. All birds to have rose combs, which, with the faces and ear lobes, were preferred purplish, or what is now known as gipsy-faced; very short-backed, short-legged, compact little birds, with drooping wings, very prominent chests, the head thrown back and almost touching the tail, which should be carried sufficiently high and open to show the clear ground colour, and distinct lacing of every feather; all the feathers, body, tail and hackles, short, and perfectly "hen-feathered" in each sex.

This club, still in existence, is, we believe, the

oldest Poultry Club in England, and the meetings are held in the Gray's Inn Coffee House, Holborn. Admission to the club can only be obtained by ballot, after having been proposed by a member, and the numbers are but small and select. Formerly, *i.e.*, a few years since, not at the commencement of the breed, there was a desire to have white ear lobes on Sebrights, but we look upon that as a great mistake. It is not only unnatural, but we may add impossible, to get a pure white ear lobe, along with a good gipsy face, on one bird. If red combs and faces were considered the correct thing, they can easily be obtained, and white lobes along with them. Or if dark gipsy combs and faces, and sky-blue ear lobes, like those of the Silky, were wanted, there would be comparatively no difficulty in combining the two points in one specimen, but a good gipsy face and white lobes, like we see on Black Rosecombs, never! And in the futile attempt to get the white lobe, colour and marking of feather has again and again been sacrificed. We unhesitatingly say, we prefer a sound, even gipsy lobe, and the freer from white, the better.

We will now describe the proper colour of the Gold Sebright. Both cock and hen should be exactly alike in colour; the ground colour should be that of the Jordan Almonds when just taken out of their shells. Many people call it a golden bay, but the term Almond colour is sufficiently explicit. Short feathers, and well rounded at the ends (not long and narrow feathers, such as we sometimes see), should be evenly edged or laced all round with rich green-black or beetle-green, and this lacing should begin on the crown of the head, and extend right to the tail end, and the hock joint. It is most difficult to get a sound tail, and next to that, good lacing on the chest and up to the throat. In the cocks especially, but to a lesser extent in the hens also, another very weak place is the wing end, in other words, the flight coverts, and to get every feather here well laced is a difficult task, and hardly any two-year-old cocks are good in this particular place. Where there is good lacing under the throat, and on the breast, there is frequently too much black on the back, shoulder, and thighs, almost always a "cloudy tail," *i.e.*, the underside of the feathers dark or grizzled.

Silver Sebrights should be pure milk white or silvery white, with every feather distinctly and evenly laced with rich green-black. The lacing should not be broader on one part of the body than it is on another, but just the same breadth all over, and in the most perfect specimens we do frequently find it so. All Sebrights should have slate blue legs and feet, but we have several times bred beautifully-marked birds with willow legs, and we have seen willow-legged Sebrights win, even at leading shows. Now, in order to breed the most perfectly-laced Sebrights, the same rules apply both for cockerel breeding and for pullet breeding, simply because their markings are alike. But from this it does not follow that all we have to do is to mate the most perfect specimens together, and set their eggs. If that were done, the result would be weaker and weaker lacing, until, eventually, it might be lost. If the hens to be bred from are fit "to win anywhere," the cock mated with them should be too heavy in lacing, and too dark for successful showing, or if the cock be a perfect show bird, his wives should be heavier in lacing, and by this means the happy medium is best secured.

One very successful breeder, a personal friend, has told us that on two or three occasions he has crossed with a Black Bantam, and then used a son of the first cross, not only to improve colour of lacing, but to increase vigour and fertility, and he has found success attend his efforts in this direction, very quickly regaining the hen feather; but we fear the jaunty shape, and nervous, tremulous motion of the Sebright has suffered consequently.

Latterly, and especially at the Dairy Show and Palace, the last two years, we have seen some Gold Sebrights of much too dark a ground colour, but their lacing was excellent. If these high-coloured birds are mated together for breeding, we should not expect many of the chickens from them to be of the desired Almond ground colour. It used to be considered a good plan, and in our own experience we proved it to answer well, to mate the very high-coloured Golds with Creamies, and thus tone down the harder colour to a softer golden hue. At one time, which we well remember, a pure silver ground was for years never seen in the show pen; and it is only about twenty-one or twenty-two years since the pure silver ground was revived again. Previous to this, our Silvers were really only Creamies of lighter shades. It is still one of the great difficulties in breeding

Sebrights that so many of the cocks are sterile, or partially so, and during the present year we have had a great number of letters asking advice on this point. Dozens of eggs have been set, but all clear and unfertile. Sometimes good and judicious feeding of the head of the harem alone, his wives being shut out of the house or runs for the time, has had the desired effect; but at other times, it has been necessary to change the cock bird. Usually it has been the best show bird that proved no breeder. Probably he was chosen because he was so very perfect in tail; but if there should be another cockerel, brother to him, with a slight inclination to be sickled in the tail, he will probably prove to be a valuable breeder, though not so good a show bird. This has been generally the experience of breeders all round.

Now, with a few hints on mating, we must close this chapter. Sebrights being an especially delicate breed, it is not advisable to run more than three hens with one cock; often two will be found to do better, where chickens are wanted; and it is safer to depend on hens from eighteen months to three and a half years old, mated with a young cockerel, than with an older cock; although we have known cases of cocks three or four years old breeding very well indeed. Such are, however, exceptions to the general rule.

Sebright's eggs may be set at any time of the year, but April, May and June are as favourable months as any to have the chickens hatched in. They require no more attention when hatched than other breeds of Bantams; liberal feeding for the first six or seven weeks will not be lost upon them, as when they get a good start in life, they do better all the way through.

CHAPTER IX.

BLACK AND WHITE ROSECOMB BANTAMS.

THESE have been general favourites for as long a time as poultry shows have been in existence, but they have very greatly improved in recent years, both in richness of colour, quality, and quantity of feather, red faces, and purity and substance of ear lobes.

There is no doubt that all these improvements are traceable to the cross with Black Hamburghs. Probably no one will dispute that Mr. E. Hutton's claim to be, at the present day, the oldest and largest breeder of Black Rosecombs is fairly established, and that to his yards can be traced the origin of nearly, if not quite, all the good Blacks now bred.

Not many breeders have kept true, like Mr. Hutton has, to his earliest favourites for a period of over forty years. But he was quite "an old hand" at breeding them, when thirty-five years ago, we paid our first visit to his yards, to purchase a pen each of Black and White Rosecombs.

Mr. Hutton, at that time, frequently bred Blacks and Whites together, and this chiefly to get better ear lobes on the Blacks, for in those days the Blacks had chiefly red lobes, and the Whites much whiter lobes, but thinner and smaller than good birds of the present day possess. One natural consequence of this cross was, that a good many of the Blacks had white or whitish legs, and a good many Whites had blue or bluish legs. The Blacks, too, in those days, were not at all reliable, in breeding, to produce sound black cocks—a large number would be black enough until fourteen or sixteen weeks old, and then would begin to show red or straw feathers in hackles and saddles. The majority were also gipsy-faced, and especially the hens; but by degrees, through careful selection, and specially by the introduction of the Black Hamburgh cross, the bad points have decreased, and the good points increased in much higher ratio, until now Black Bantams are very near perfection.

In the early days, to which we just now referred, there was much difference of opinion as to whether long legs or short ones, long and drooping wings, or short and high carried wings, should have the preference; but here Mr. Hutton took a decided stand from the first in favour of short leg, and the jaunty carriage, quite opposite to the style and carriage of Game Bantams; and though now very slightly modified to resemble miniature Black Hamburghs, the majority of judges and breeders have agreed with the correctness of Mr. Hutton's selected type through all these long years. The ideal Black Rosecomb Bantam now, is a Black Hamburgh on a small scale, and about one-fifth the weight of the Hamburgh. In feather, both cock and hen should be as green a shade as possible; the combs and faces, coral red; lobes, white and soft as a fine kid glove; legs, black; the cock's weight from 18 oz. to 22 oz., hen's weight 14 oz. to 18 oz., will be found about the correct actual weights.

In breeding Black Rosecombs, it is still found that the best results can be obtained by keeping one strain to produce cockerels, and another to produce the pullets; for it is almost impossible to breed the rich beetle-green pullets, with coral red combs and faces, and good white lobes, without the cockerels coming red in hackle and saddle, and on wing; and equally impossible to breed sound, rich black cockerels, except from hens so dull black in colour, as to stand a poor chance of a prize in good competition. It is better, therefore, to select two sets to breed from, as advised in the case of the Black Pekins.

In selecting a breeding stock of Black Rosecombs, special care must be taken over the combs of both parents; hollow centres and faulty peaks, comb short of "work," *i.e.*, spikes, must be carefully avoided, for half the battle lies in the head points, and faults are more persistently reproduced than excellencies. Ear lobes must also have a fair share of attention, and as much in the hens as in the cocks; quality of lobe must be sought for, and maintained. We do not recommend a larger number of hens than three or four to be mated to one cock, and advise that most of the eggs be set in March, April, and early in May, as the chickens hatched at this period of the year feather the best of any; this especially applies to the cockerels.

Many people hatch late in the year, so as to keep the chickens small. This plan answers well enough for hens, but not for cocks, as they do not feather up well, and, therefore, are useless for many months, except for breeding.

When the chickens hatch, all the good ones will be black on the upper parts, with white tips to the wings, white throats and bellies, and more or less white on face.

Should any be entirely black when hatched, they will not be found black when in full feather, but show more or less red or straw feathers. Black Bantams are good layers, and many lay from 100 to 150 eggs per annum, weighing about ¾ oz. each, or about 11 or 12 to the pound. The competition for the challenge cups is every year an exciting event at the Crystal Palace, and one of the most successful competitors is Mr. E. Wright, who has bred and exhibited some of the best Black Rosecomb cockerels and pullets ever shown. Another dangerous rival is Mr. E. Walton, and several others follow closely on their heels.

White Rosecomb Bantams have recently improved quite as much as Blacks have, and it would be difficult to conceive a more perfect specimen of the White Rosecomb cock than the one recently exhibited by Mr. E. Walton, and, of course, always winning. The Whites should be pure white all over the feather, the beak, legs, and lobes; the face and wattles, coral red; and the feather as abundant as possible. The cock's sickles, and side sickles, a good length, and indicative of the bird being in full vigour. The eye should be a clear bright red, and fiery looking.

If all these points are manifestly present, the birds have a good chance of beating equally good Blacks in any competition, but it is seldom that we see all good points combined in one bird; very frequently, the otherwise best cock has got tanned by the sunlight, and his hackle and back are quite yellowish coloured. Others would be capital birds, but their legs have a decided bluish tinge, or their lobes are not thick enough and white enough. There is more trouble in getting a good White cock than in finding two good White hens, though very few of these are quite as good in lobes as they should be. In this point, our American cousins have decided that they must have coral red lobes on the White Rosecombs; and certainly the red is a more striking contrast on the pure white feather, and looks very well, quite as well as the white lobes look on a black background of feathers.

No different breeding pens to produce the opposite sexes are required for White Rosecombs. Purity of white, accuracy of shape and texture of lobes, and of shape and quality of combs, must be carefully sought for, and the birds must be kept shaded from sun, and sheltered from rain and cold winds, or else they will be spoilt for showing; but this is no more than is necessary for all pure white breeds. One of the most difficult points to obtain has been good full tail furnishing, except along with bluish legs. It has, however, been obtained in Mr. Walton's best birds; and what has been done, can be done again. Therefore, let other White Rosecomb breeders say "Excelsior," and they can succeed.

And now, in conclusion of our remarks on Black and White Rosecombs, we wish to enter a protest against what we always have thought, and still think as strongly as ever, is a great mistake in breeding these pretty birds, viz., the attempt to get them as small as possible. We believe it is possible to breed them so small that, perhaps, they could be shown full grown, the hens not weighing more than four or five ounces, and the cocks half a pound, but *cui bono?* What *use* would they be? The experiment may be a curious one, and interesting to some minds, but for practical purposes, and for exhibition, we feel convinced that the most satisfaction all round will be given and secured if we observe the happy medium, and content ourselves with birds averaging one-fifth the weight of the large varieties which the Bantams represent in miniature; that would be about 16 oz. to 20 oz. for Black or White Rosecomb cocks, and 14 oz. to 17 oz. for hens, in good show form. And we should be glad if a common-sense rule were applied, deducting so many points for too small size, in the same ratio as for too great size. We say this deliberately, after more than forty years' experience in Bantam breeding, that when we produce them below such a point of size as the one we have named, we immediately begin to impair their constitutions, to ruin their health and vigour, and consequently their beauty, and to lay the foundation for our own, and other fanciers' disappointment with the breed.

When, in 1871, we were asked to contribute an article on Game Bantams for the "Illustrated Book of Poultry," by Lewis Wright, published by Cassell and Co., we gave the weights for Game Bantams as follows:—Cocks not to exceed 26 ounces, nor hens 20 ounces. It was then a rare occurrence to find Game Bantam cocks carrying their wings clipped up to the body, as we now see them, and it had been necessary to cross, again and again, with large Game fowls to improve style and carriage, so that a little more weight was then usual, and had to be allowed, than now, after another twenty-one years of careful breeding; and we should be quite prepared now to reduce those weights by four ounces in the cocks, and three ounces in the hens. This would bring them down to the proper standard weights, below which we would not try to go. But there is one point in the editor's footnote which we cannot agree with, and as it applies to all Bantams, we here give the quotation *in extenso*, and our reasons for differing from Mr. Lewis Wright's conclusions on the point:—Page 488: "We have found, with birds in average condition, that 22 ounces in cockerels, and 18 ounces in hens, are what will fairly entitle birds to be called 'perfect,' and about one point should be deducted for the first ounce over this, two points the second ounce, three points the third ounce, and so on, *while two points per ounce may be credited for less weights*. But, as already observed, individual birds differ much, and it is the *apparent size usually denoted* by these weights, and not the weights themselves, that are to be considered by the judge."

Mr. Wright's standard weights, it will be observed, twenty-one years ago, are what we are fully prepared to adopt now for cocks (not cockerels) and hens, but they were then below actual weights of the best winning birds, and were rather looked upon as ideals to breed up to, or, as some would say, to breed down to; and with his handicapping, by penalties, excess above these weights, we find no fault, but having chosen an ideally perfect bird, we do consider it a mistake to credit every reduction of an ounce by two points gain over the ideally perfect specimen, and we know that this idea has tended to the ruin of more than one fancier's birds, who have tried to reduce size of cocks to 14 oz. or 15 oz., and hens to 11 oz. or 12 oz. True, they have got the small size and weight, but their birds are puny, delicate little things, and very difficult to keep in health.

We would advise judges, exhibitors and breeders to be satisfied with a reasonable limit in size and weight, and neither try to breed, nor encourage, smaller or larger specimens than that reasonable standard. What has led us to make these remarks at such length is the fact that we know some exhibitors who have, over and over again, quoted Mr. Lewis Wright's note here referred to, to back up their claims to win with absurdly diminutive specimens of Bantams, six or eight ounces in weight, that were perfectly useless for reproduction.

CHAPTER X.

CUCKOO OR SCOTCH GREY BANTAMS.

AS we stated in our opening chapter, we believe this to be a comparatively recent production, but whether English or Scottish breeders rightly claim to have originated them, we cannot find out. We remember seeing them some thirty years ago, and those were undoubtedly of Scottish breed.

But as we have known of them for about the same length of time in the South and West of England, and also for many years in the Midlands, we will not pretend to say who first bred them. It is a well-known fact among breeders of experience that the cuckoo marking can be produced by the fusion of a pure black with a pure white breed, and no doubt Cuckoo Bantams were at first so produced, and very possibly, simultaneously, in various parts of the United Kingdom. We know that Rosecombed Blacks and Rosecombed Whites were crossed for the purpose of improving the ear lobes of the Blacks, and in all probability the Rosecombed Cuckoos are a sport therefrom. But whether or not that be so, we have the two breeds, and so, without further enquiry as to their source or origin, we will proceed to describe them as they are, and to treat upon their mating and breeding.

Of the two varieties, the single combs have the preference; and the aim of all breeders seems to be the production of the Scotch Grey fowl in miniature.

In size, Cuckoo Bantams should be small; as small, in fact, as Black and White Rosecombs. The combs must be perfectly erect in both sexes; ear lobes, red; combs, wattles and faces, red; legs, white, or slightly mottled to match the plumage; eyes, as red as possible; feather, finely and evenly marked with distinct bars or bands across each feather, and the finer these bars, the better. The ground colour should be a very pale French grey, both in cocks and hens, and the bars should be of a dark slate colour. Clearly defined colours and markings are highly desirable; and no white feathers, in either wings, or tail, or elsewhere are admissible; black feathers are often seen in the hackles and saddles of cocks, but they are not desirable; at the same time, they are not so objectionable as white.

It appears to be quite a common practice for Cuckoos to leave at home such objectionable feathers, when they are about to set off for exhibition; and one old breeder of this variety, in reference to this point, once said in our hearing: "I would not give a fig for any bird that wanted no trimming. Why, man, half the pleasure of showing consists in making the birds look their best before the judges."

The late Mr. Phelps, of Ross, was for years one of the most successful breeders and exhibitors of them, and then Miss Jackson, and Mr. Edwin Wright. They were also bred in great numbers at Old Windsor, and in various parts of Nottinghamshire. The colour of the Cuckoo is, however, so very susceptible to the effects of sun or rain, and is so very fleeting, that unless good shade is provided—as for white fowls—they are soon quite spoilt, the dark grey bands becoming brown, and the light ground colour quite brassy. In this state it is useless to show them.

In breeding Cuckoos, we always prefer a cock bird rather darkish in colour, and as fine in bars as possible. The natural tendency of the breed is to revert to the broader and opener markings, which are of less value. Occasionally, it is necessary to breed from black hens, in order to strengthen or intensify the colour, but only those black hens should be used that are Cuckoo bred, at least on one side. At times, white chickens sport from them, but they are of little or no value, and we should never breed from them ourselves.

Most of the Rosecomb Cuckoos we have seen

had ear lobes far more white than red, though red is the correct colour for the lobes. ' As the leading judges gave the preference most decidedly to the single combed, we believe the rose combs have been allowed to die out, or very nearly so, as we do not remember seeing a pen of the Rosecombs for four or five years, and those were in Nottinghamshire. It would, however, be a very easy thing to reproduce a strain of Rosecomb Cuckoos, by mating a well-marked single comb Cuckoo cock with Rosecomb Black hens, and then mating a Cuckoo cockerel bred from them to the Rosecomb Black hens again, until the desired result was produced. Cuckoos are tolerably hardy, and about as good egg producers as any Bantams. They also make good sitters and mothers.

Occasionally, Cuckoos are seen with yellow legs and bills; but those are not desirable points, unless it were attempted to make miniature Plymouth Rocks, which would be a very easy task, though we doubt if it would repay anyone to undertake it.

We have already a very elegant and well established breed in the Cuckoo Pekin previously described, and we have also produced Cuckoo Polish Bantams, which came uninvited, while we were making the White-crested Black Polish some few years ago.

BLACK ROSECOMB BANTAMS.

Bred by, and the Property of, Mr. Edwin Walton, Horncliffe, Rawtenstall, near Manchester.

Cock, winner of First and Cup at Otley, Clitheroe, etc., 1893.
Hen, First and Special at Birmingham, Firsts at Leeds, Liverpool, Durham County, etc., 1893-4.

CHAPTER XI.

NANKIN BANTAMS.

THESE, we believe, are amongst the very oldest of our English varieties, and our recollections of them extend to more than forty years back, when we much admired a pen of them belonging to a neighbour. These were Rosecomb Nankins, with bright slate coloured legs. There are also single comb Nankins, some of which are white-legged, and others blue-legged. Nankin Bantams we believe to be the foundation from which many other breeds have sprung, including Sebrights and Game, and it would only take a very short time to produce Golden Pencilled and Silver Pencilled Hamburgh Bantams, by using Nankin hens with the Hamburgh cocks of those varieties.

Though, undoubtedly, a very old breed, it appears unusually susceptible to impressions from almost any breed crossed with it, and has, therefore, been so useful in the production of other races.

The proper colour of the Nankin cock is a darkish cinnamon colour all over, with the exception of the flights, which are chiefly black, and the tail, which is bronzed, though chiefly black. The hen, also, is similar to the cock, though one shade lighter in body colour; her tail is almost black, the top outer feathers being chiefly cinnamon, to match the body. Of late years, this variety has lost its popularity, and is becoming more scarce each year. Mr. Cresswell and Mrs. Ricketts have been foremost in their attempts to preserve the breed from utter extinction, but we are afraid it may be a long time before there are many pens seen at our shows, unless a class (or classes) are guaranteed at the Palace, or some leading fixture.

If those breeders of Nankins, who desire to see their pets become more popular, will only communicate their wishes either through the Bantam Club, or direct to the secretary (Mr. Howard) of the Crystal Palace Poultry Shows, and will guarantee the show committee against loss, there is no doubt that classes can again be added to the already magnificent schedule; and this, more than anything else, would tend to increase the interest in the breed, necessary to its extension in fresh quarters. And with regard to its delicacy of constitution, that may easily be overcome in a variety of ways, but first and best by an interchange of stock of pure blood, as now existing. This might easily be accomplished through advertisements in the Poultry journals, which we have reason to know are extensively circulated among, and read by, Bantam breeders in all parts of the world. An advertisement repeated for a few weeks should result in bringing in the names of most of the Nankin breeders to one centre, so that either a club could be formed, or exchanges of birds be made, with the result of strengthening the breed all round. A few years ago, when Buff Pekins in England were fast dying out, two of the then remaining owners of stock tried the cross with Nankin Bantams to increase fertility. This it certainly did, but it also quite spoiled the shape and feather of the offspring, which even, when bred back to Pekins, produced very un-Cochiny looking chicks.

If a cross of fresh blood should be deemed advisable for the Nankins, it might be obtained from either the Wheaten Game Bantam hen, whose Nankin crossed pullets should be bred back to another Nankin cock, and their pullets again to a third cock of the same breed as their sire, destroying all other chicks of the first and second crosses, and retaining pullets of the third cross to use as future hens for breeding Nankins. This would be our plan; but either Black or White Rosecomb Bantam hens, or even small Gold Pencilled Hamburgh hens, might be used instead of the Wheaten Bantams, the same plan of operation being adopted with their offspring.

Suitable fish diet, of which the most convenient

form is the Liverine meal, tends very greatly to increase the fertility, as well as the general stamina, of delicate fowls, and will be found invaluable in building up a fresh strain of any worn-out breed. Another valuable adjunct in the rearing of delicate chickens is "Parrish's Chemical Food," or syrup of phosphates, which may be mixed, in small quantities, of course, with the meal with which the chickens are fed: two drops a day to each chicken for a fortnight at a time, and then discontinued for a fortnight, will do good.

In trying to revive Nankin Bantams, we advise everyone to endeavour to perpetuate *blue* legs, and discard all other colours. The blue legs are natural to the breed, and there will be a stronger tendency to revert to blue if any other colour is tried for, than to any other shade, when the desire is to breed only blue legs.

One more word of caution before we pass away from the subject, and that is: Don't try to breed Nankins so ridiculously small that they cannot possibly maintain their vitality. Let 16 oz. for hens, and 19 oz. or 20 oz. for cocks over a year, be considered quite little enough weight; then they will once again have a chance to live and do well.

CHAPTER XII.

JAPANESE AND FRIZZLED BANTAMS.

BOTH of these varieties are comparatively new introductions to our country, and the Frizzles are the newer of the two. Since the commercial intercourse between Japan and our own country has been opened out, we have received from time to time many consignments of Japanese fowls, Bantams of various colours and kinds, and those very curious long-tailed birds, the Shinawaratas, the Shirafusi, the Yokohama, etc., many of which are but little larger than Bantams, and yet have such enormously long tails, and abundant ribbon-like sickles and streamers. These must have taken ages to have produced, and as crossing with other breeds would utterly ruin the point of most value —length of tail—an amount of in-breeding must have been resorted to, which is utterly opposed to general ideas on the subject in this country.

However, the Japanese Bantams which we have now to consider are almost exclusively of one type, varying only in colour. It is about thirty years since the first pens of these birds were imported into England, and they were of the well-known Dark-tailed, White-bodied variety, most frequently met at our leading shows. They were at first hard to acclimatize, and many died during their first winters in England, but by degrees they became more used to our varying climate, and may now be classed among the half-hardy breeds. For some years, we never saw such perfectly edged sickles in this variety as have been shown at the Palace, Birmingham, and our other large shows during the last six years; but we have been more exacting in our requirements when importing—we have more clearly defined standards to judge by; and, undoubtedly, we have developed, in this country, such points as we aimed to secure. The result is, that now we can find in a first prize pen at a leading show, the cock bird, with a snowy whiteness of feather in neck, breast, thighs, body, and wings; and with a black tail, every sickle feather of which is evenly-edged with white; his legs and bill are as yellow as an orange; his comb, lobes and face as red as blood. When his wings are open, they show the black in the underneath feathers of flights and coverts, but the upper and outer surface is white; the tail is carried very erect, the more erect the better—either a squirrel tail, or the other extreme, a drooping tail, is considered a grave fault. The hen is coloured exactly like the cock, and her shape should be like his. The comb is single, large, and well serrated in all these Japanese.

We have found them breed very true to points, and cannot give any special directions for mating, further than to advise care in selecting clearly edged feathers on as sound a black ground as possible, the chief tendency towards degeneration being greyish tail feathers, and indistinctness of lacing.

Besides the above, which are the commonest, there are other colours now met with, newer importations, and offspring from such, viz., solid Blacks, solid Whites, Buffs, also Greys and Browns of various shades; but of all these, the Dark-tailed Whites and the solid colours will, and do, take the lead with English fanciers. Whatever the colour of the feather, the legs should be a bright orange yellow, and the beak must also be yellow.

Japanese Bantams look larger than they really are, on account of their length and abundance of feather. In shape they are peculiar, and unlike most varieties. They have exceedingly short shanks and thighs, so short that their bodies almost touch the ground; their wings are long and broad, and carried with the points downwards; the head erect, breast rather prominent, and tail perfectly erect. Weight: Cocks between 18 oz. and 22 oz., and hens 14 oz. to 18 oz. The best time to hatch their eggs is May, or early in June, then the chickens grow very quickly, and soon acquire their first suit of feathers, when the are very attractive little things.

The Frizzled Bantams are closely allied to the Japanese, but very little appears to be known clearly as to their origin. At the present time, very conflicting statements are published and circulated in reference to them. Among the most successful exhibitors of Frizzles, from the earliest recollections we have of them, are Mr. F. C. Davis, Mr. Cox, The Countess of Dartmouth, Mrs. Ricketts, Mrs. Frew, Mr. E. Morgan, Mr. Billett, Mr. Lawther, Mr. House, etc. Speaking from memory, it is about twenty years since we first saw Frizzled Bantams exhibited, and they were Whites, but since that time, we have seen at the Crystal Palace Show, White, Black, Gold, Brown, or Partridge, Dark Grey, and Blue or Slate-coloured Frizzles, and all well curled. What was the origin of this peculiar curl of feather we cannot say, but it is not confined to Bantams; many large breeds also have their frizzled sub-varieties, and we have heard and read of Frizzled Cochins, just as we have heard and read of Silky Cochins, and Silky Langshans. Their peculiarity consists, as in the Frillback Pigeon, of all the feathers being curled backwards, and turning back towards the head, and the sharper and closer the curl of every feather, the more perfect the specimen. The Frizzles vary very considerably in other points. There are single combs, rose combs, and cup combs; clean-legged, feather-legged, four-toed, and five-toed, and we have even seen birds varying so greatly, exhibited in one pen, and winning. In fact, on one occasion, when we were at a show, the judge who was officiating appeared much struck with a trio which were shown in the variety class, and, singularly, one was four-toed, clean-legged, and single-combed; one five-toed, feather-legged, and single-combed; while the third was four-toed and clean-legged, but rose-combed. The judge awarded them the first prize, and when questioned by another exhibitor in the class as to which of the three was the correct type, and how it was they had won, he answered, "Why, all three are right for the 'variety class,' and it is the only 'variety' pen in the class, and that's why they won!" It was this state of things that induced Mr. F. C. Davis, and those friends who could not appreciate the force of such logic, to make a move towards securing greater uniformity of judging, by drawing up a standard, which was printed and published in one of the fanciers' papers; we forget which. But we do recollect that one of the judges at the Crystal Palace, whose name was down for the variety Bantam classes, refused "to be bound by that or any other standard," and consequently entries were withheld that otherwise would have been sent to the Palace.

We, personally, have no *great* preference for the single comb over the rose comb, nor *vice versa*, in this breed, though we may own to a slight leaning towards the single combs; but in the matter of legs and feet, we say, decidedly, four toes and clean legs are to be preferred; and that, unquestionably, all birds shown in one pen must match each other in such points. Of all the Frizzles, Whites seem to be the most charming. These should have yellow legs, but very often their legs are willow, sometimes slate, but yellow should have the preference. Next come the Golden with yellow or willow legs; the Slate coloured, with black or slate legs, and the Black with black legs, then the Browns, Greys, etc., with dark legs. We have seen them bred in large numbers by Mr. Davis, of Southampton, and from his stock most of the good birds of this variety have, no doubt, sprung either directly or indirectly. Frizzled Bantams are very small, many winners not exceeding 1 lb. in weight; they are longer in shank and thigh than Japanese, but length of leg is not a point of much note; the most valued property is curl, then quality of feather, hard and wiry, then colour, next size.

They are very susceptible to changes of weather, and should never be allowed out in a shower of rain. Like Japanese, they are excellent sitters and mothers, but should never have more than six or seven of their own eggs to hatch.

JAPANESE BANTAMS.

Bred by, and the Property of, Mr. O. E. Cresswell, Morney Cross, near Hereford.

Dark-tailed Light Cock, winner of Third Prize at Crystal Palace, 1893.
Black Hen, "Dorothy," Firsts and Cups at Crystal Palace, etc., etc., 1892 and 1893.

FRIZZLED BANTAMS.

Bred by, and the Property of, Mr. Geo. Reyner, Thurlstone, near Penistone.

White Cock, winner of First and Cup at Crystal Palace, First at Birmingham, etc.; and Black Hen, winner of First at Crystal Palace, and Reserve at Birmingham, etc., 1893.

CHAPTER XIII.

POLISH BANTAMS.

POLISH Bantams are simply miniature Polish fowls, and are of various colours, viz., Black, White, Buff, Gold, Silver, Creamy, Cuckoo, Blue, White-crested Black, and White-crested Blue. These are the ten colours we have bred, but we fear that some of them have already been lost, and allowed to die out, although we know that there are fanciers still breeding a few each year of some of the colours.

In commencing to breed Polish Bantams, we endeavoured to eradicate all traces of comb and wattles in the least possible time, and chose the produce of our various matings of stock for head points, in preference to colour or markings, which we expected to fix with less difficulty. We first visited the yards of three or four breeders of the large Polish fowls, and at a small outlay, secured several very large crested birds, that were of too small a size to please their breeders; explaining that our only object was to breed them still smaller, and, in fact, to make Polish Bantams as soon as possible.

We also obtained a few Sebright hens, both Gold and Silver, and also White and Black Rosecombs, and at once set to work to obtain the necessary crosses. In each case, so far as possible, we carried out our crosses on the well-known theory, that the dam influences size, and inward characteristics to a greater extent than the sire does, whereas he, in turn, should stamp his impress more plainly on the outward form and colour of his offspring.

In the case of Gold and Silver Polish, we never had the least trouble with regard to colour, for we mated a very small Gold Polish cockerel with an extraordinarily large, but wild, crest, with Gold Sebright hens; and, without having to wait long, we had a large number of fertile eggs, from which we raised quite a number of big Bantams of lovely colour and markings, and with small crests and combs; the pullets were in turn bred back to their sire, their offspring coming with a very fair amount of crest. Of course, it will readily be seen that each cross to improve crest and eradicate the comb, or tendency to comb, naturally increased the size of the birds; but as we obtained two crosses each way, in each season, we were enabled to win first prizes, and a silver cup, with Gold Poland Bantams at the end of the sixth year of breeding; and those birds were then as perfectly free from comb, and as large and globular in crest, proportionately to the size of bird, as the winning large Polands were. With Silvers, we adopted precisely the same mode of procedure, only using Silver Poland cockerel and Silver Sebright hens together, and when we obtained the produce, we mated the largest crested small cockerel with the Sebright hens to breed pullets from, and the first cross pullets back to their sire, to breed a cockerel or two to mate with the smallest pullets, and so we bred in-and-in, and, by careful selection, soon established a reliable strain. To obtain self-Whites and self-Blacks and White-crested Blacks, we placed both White and Black Rosecombed Bantam hens, along with very small White-crested Black Poland cockerels, and though we cut the crests of the cockerels, we had to try three or four before we could succeed in getting any fertile eggs and chickens. At last, and not till after we had entirely wasted two years in fruitless attempts, we had a lot of chickens hatched, but there was no approach to anything like a White-crested Poland among them. There were some all black, others all white, some speckled, grizzled, and even Cuckoos in the first cross.

We, however, persevered, totally ignoring the colour, and considering only crest, and head points generally, and size; and not until the fourth and fifth cross, had we anything encouraging in the matter of colour in the direction of White-crested Blacks—they were pullets.

In course of time, however, more and more chickens came with fair-sized knobs, which were quite white when hatched, and as we looked them over carefully, we hoped that, at last, they were coming right; but for three or four years, even these best looking chicks, with pure white knobs, as they grew up and changed the first head-gear for the second, would develop, here and there in their crests, too many black feathers. No doubt, they could have been trimmed and made to pass muster; but we never showed one that was faultless, each one had some black in the crest. Whilst we were trying for the White-crested Blacks—the White-crested Blues, the self-Blues, Blacks, and Cuckoos came unbidden, and uninvited, and, at the same time, we were making rapid progress with pure Whites, which had ample crests, whiskers, and beards, and all of them blue legs, and several no comb at all. With these Whites we won many prizes, including cups at the Crystal Palace, and Manchester, and other shows. One of the best of these was a White hen, which, in her second year, was only fourteen ounces in weight, and a most perfect little miniature Poland hen. She went, along with the rest of our Poland Bantams, to Magdeburgh, when we gave up the breed in November, 1888. Since then, she has twice crossed over to England, to compete with our best productions. These Whites were totally distinct, and different in their origin, from a breed of small White Poland Bantams, existing at the same time in the United States of America, of which we were in total ignorance, until one of the leading breeders there suggested to us that an exchange might be mutually advantageous.

We agreed, and made an exchange of a trio, and we sent our blue-legged, crested, and bearded Whites, receiving a trio of very tiny birds, less in size than ours, but with combs which frightened us by their size, and, also, they had pure white legs and feet—another point we thought undesirable. We found these little hens very useful, and good breeders when mated with one of our own blue-legged White cocks; but the little American White cock never bred us a single chicken, and we soon gave him away. The produce of the little American hens needed breeding back again, a few times, to cocks of our own strain, before we could find any of their progeny without combs, and with the proper-shaped knob, from which a full and globular crest must grow; and it was some time before we eradicated the tendency to split crests. There was more difficulty in producing pure white cockerels than pullets—either straw-coloured or sandy-backs, or a few coloured feathers often appearing where they were not wanted; but sound white pullets came in numbers.

In the case of Blacks, too, the only difficulty in the way of colour, occurred with the cockerels, which generally grew some red or straw-coloured feathers in the hackles, saddles, and wing-bows, and sometimes a few red or other coloured feathers in the crests. A few odd birds that we bred were very pretty, though peculiar. There were some Red Pile cockerels, and one or two pullets, some with rich creamy-buff ground colour, every feather tipped with a small green-black round spangle, and some almost grouse coloured; and, strange to say, these odd coloured birds were about the best crested we bred, and most of them turned out very useful in breeding the standard colours. We freely parted with our Polands, right and left, not caring to confine the breed to our own yards, but rather wishing to see them become an established breed in the country; and we are glad to know that several fanciers are still breeding them: Whites, Buffs, and White-crested Blacks chiefly.

Buffs were, at first, sports from Golds; but most of the Buffs had some little admixture of either white or black in their feather, and we never set ourselves the task resolutely to persevere until we had bred a solid buff, even in colour all through, though we do not think the task impossible of accomplishment.

The Blues and Cuckoos both came during the crossing of Black with White, in the production of those colours, and the White-crested Black; and one season, we raised a lovely trio of White-crested Blues, until they were nearly full-grown, when they were killed and carried away by rats, not, however, until they had been seen and admired by several fanciers. There would not now be much difficulty in the way of fanciers breeding any colour of Poland Bantams, as materials exist, though, perhaps, somewhat widely scattered about the country, which are the result of many years breeding, and only require judiciously mating to result in a good strain being perpetuated.

We only need to add that the recognized standard for the large Polands should be regarded as the correct standard for all Polish Bantams, excepting, of course, the size, and that one-fifth the usual or average weight of the Polish fowls be taken as the standard weight of the Bantams, *i.e.*, 14 to 18 ounces for pullets and hens, and 17 to 22 ounces for cockerels and cocks. One word of caution we perhaps ought to give to all breeders, and that is, always select your breeding stock from those birds which, as chickens, show the largest knob, or cushion, for the crest to grow from. Those chickens whose knobs are small, or only grow too far back on the head, are never worth breeding from, and the best thing to do is to kill them while young—only, the inexperienced fancier doing this might not improbably kill off *all* the pullets, and have only cockerels left, their knobs being larger than the pullets when hatched; and we would never breed from a cock, or cockerel, that had any comb further than the two little horns, which are always more or less in evidence, but always select those with good frontal crest, and well-filled centre, nicely radiating to the circumference.

We hope to again some day see classes established at the Palace Show for Polish Bantams, as during the time when we guaranteed the entries ourselves.

CHAPTER XIV.

MINORCA, ANDALUSIAN, LEGHORN, AND HAMBURGH BANTAMS.

NONE of these have, as yet, made any real headway in public estimation, excepting the last named, and of those only the Black and the White, really miniature Hamburghs, though commonly known as Black Rosecomb and White Rosecomb Bantams. As these two colours were described in Chapter IX., we will now proceed to consider the other colours, Spangled and Pencilled, and though we have already bred one or two specimens of each colour, that have been described as "very good attempts at," yet, that is about as far as we have gone; and, unfortunately, both time and space failing us, we parted with the whole breeding stock. We must, therefore, treat of these more as a possible future breed, and rather indicate the way in which we should proceed to breed them, were we about again to start on that task, than consider them as an existing variety. We may, however, say that in both Gold and Silver Spangled, and Gold and Silver Pencilled Hamburghs, there exists such an amount of prepotency, owing to their careful breeding through many generations, that, when crossed with other breeds, which I shall have to name as suitable for the purpose, the Hamburgh markings are easily reproduced. First, then, let us take the Gold Pencilled as the smallest of the Hamburgh tribe, and, therefore, the nearest to Bantam size, to start with. The plan would be to select a very small, slender-boned cockerel, of the finest Pencilled pullet strain, and mate him with two or three tiny little Rosecombed Nankin hens, with blue legs, and rear, say, a score pullets and cockerels from them. When old enough, select the three best marked, small-boned pullets, and mate them with their sire, at the same time mating the most Hamburgh-like cockerel to the Pencilled hens, and in the following crosses, always selecting the most typical representative of the Pencilled Hamburgh cock, as the cockerel to breed from, looking particularly at comb and lobes for the cockerel breeders, and at comb, lobes, and pencilling while the cockerels are in first feather, for the pullet breeders; and in each case choosing the very smallest pullets possible, which show the desired Hamburgh points. With two crosses a year, four years' work ought to produce something deserving of a prize at a good show.

With Silver Pencils, the best birds to use would be the Silver-Pencilled Hamburgh cockerel, and (don't be frightened, and say impossible) the Dark-tailed White Japanese hens, with their yellow legs—a point soon exchanged for blue. We say Japanese hens, rather than White Rosecomb hens, because, from experience, we know it is easier, and a saving of at least eighteen months or two years time. The produce of the Light Japanese hen should all be killed, excepting four or five of the most suitable pullets reserved for mating to their sire, and for the next cross, one of his sons might be reserved to breed with some of the smallest pullets. The colour of the cockerels will be perfection, and the pencilling of the pullets will be improved, year by year, by careful selection. It will readily be seen, that the cross of Silver-Pencilled Hamburgh cock and Light Japanese hens ensures the perfection of colour in the cockerels' tails especially, and fairly good colour in the tails of the pullets. There will, naturally, be a tendency to faulty carriage of tail, if care be not exercised in selecting a Hamburgh cock that carries his tail low enough. The difficulty of comb and ear lobe we overcame in the second cross back to the Hamburgh.

In the case of the White Rosecomb hens and Silver-Pencilled Hamburgh cock, our first chickens were all black ones, but in the second cross, we had some that showed a fair amount of pencilling, and others distinct spangling; and when we mated the best pullets of this lot with the Pencilled cock, we had some really good Pencils, but almost large enough to show in the Hamburgh class.

We tried, also, both Gold and Silver Sebrights with Pencilled Hamburghs, and proceeded as far as the third cross, the result being larger birds than those from the Nankin, Whites, Rosecombs, and Light Japanese crosses; but, we believe that perseverance with any one of these breeds would have been rewarded with success, had the opportunity been given us to proceed farther. We would, however, emphatically say, do not on any account mix up three of the breeds in one strain; whichever you select, whether Sebright, Nankin, Japanese, or White Rosecomb to cross with the Hamburgh, work up the produce between the two breeds, without adding any third cross. The reason will be obvious to any who have had experience in crossing pure breeds.

For the Spangled Hamburghs, the very best birds to cross with are Black Rosecombs for the Golds, and either Black or White (immaterial which) for the Silvers, and the time occupied should not exceed five or five and a half years, from commencement to the time when specimens should be able to win.

We do not, for a moment, think that any pecuniary profit will accrue from the production of Bantam Hamburghs; but from the scientific point of view, there may be a vast amount of pleasure in producing a new variety, and in determining the length of time which is required for the evolution of a new species. And such pleasure lies within the reach of many a poultry breeder.

But we must pass on, and for a few moments think of the Minorcas and Andalusians. These being much larger birds than Hamburghs, have necessarily proved more tedious to "bantamise." The way in which we advised an enquirer to work, some few years ago, was the following, for he wanted to produce both Andalusian and Minorca Bantams, and being a breeder of both varieties of the large fowls, we said: "Turn three or four Black Rosecomb hens into a run, with a small but typical Andalusian cockerel, and probably some of the eggs will hatch; then select your pullets from them into two sets, and mate one set with a very large-combed Minorca cock, and the other lot with an Andalusian cock, and then in-breed with both families." He did so, and wrote us at the end of the first and second years, reporting distinct progress. At the end of the third year, we met at Birmingham Show, and talked matters over, when he told us he should be able, in another twelve months, if all went well, to place some pullets fit to win in the show pen, but the cockerels would still require more time. He then had beautifully laced Blue pullets, not exceeding 2 lbs. weight, laying. For some years the Countess of Dartmouth has had a strain of Blue Andalusian Bantams, with which her ladyship has won numerous prizes at the Crystal Palace and Birmingham Shows, but they have been rather large Bantams, and hardly so good a blue as some large Andalusians—their combs, lobes, heads and general shape have been highly typical.

Leghorn Bantams have been taken underhand by a few breeders, who have attempted to make Browns, Whites, Duckwings and Piles, and for this purpose have availed themselves of Game Bantam hens of each of the above colours, the Browns and Whites being, of course, the produce of Pile Bantams. The difference, however, between Game Bantams and Leghorn Bantams, of these various colours, would appear to be all in favour of the Game, which are smarter, and especially much neater about the heads; and we, therefore, do not, and cannot, advise anyone to take the trouble to breed Leghorn Bantams.

It will, however, be readily seen that the worst Game Bantams, *i.e.*, those with large combs, fleshy faces, whitish lobes, yellow legs, and rather bulky bodies, would be the best for producing the Leghorn points, and that such birds should be mated with the smallest available Leghorn cocks of suitable colour, and with excessive development of comb and lobes. It would then be a comparatively quick and easy task to produce winners, which would require no crossing to keep up the strains. In-breeding, for twenty years in succession, would improve rather than injure them, providing that all sanitary arrangements were right.

The size of Leghorn Bantams, and Andalusian Bantams, should be the same as that of Game Bantams. The Minorcas might be allowed 1½ or 2 ounces more in weight, and look rather larger; while the Spangled and Pencilled Hamburgh Bantams should be the same in size and weight as Black and White Rosecomb Bantams.

CHAPTER XV.

RUMPLESS BANTAMS, AND OTHER VARIETIES OF BANTAMS NOT PREVIOUSLY DESCRIBED.

THE Rumpless or Tailless Bantams are, perhaps, the most curious of all the varieties, and also very uncommon, as they are in the hands of only a few fanciers, and are rarely exhibited.

Mr. Tegetmeier was the originator of this breed. When working with the late Mr. Darwin on the subject of variation in animals, he procured from Turkey a very small White Rumpless fowl. This bird, a hen, had a small crest, and was mated with a White Polish cock, and in two or three years produced some fully-crested, Tailless Polish, which were prize winners. These were of the large variety, and weighed from 4 lbs. to 5 lbs. each.

Mr. Tegetmeier then resolved to reduce the size of the Rumpless fowls, so mated the original hen with a tiny Nankin Bantam, and had no difficulty, in two seasons, in producing some pretty little Tailless Nankins, which bred semi-wild in a wood near his house. Having won first prizes at Birmingham, and other shows, and accomplished his purpose, which was to prove the variability of the breeds of fowls, he allowed the whole stock to pass into other hands.

Rumpless Bantams are found in several forms and colours. There are some with short legs, heavily feathered and booted, and others with longer limbs quite free from feathers. They may also have either single or rose combs, though the former are more general; but all are alike in being destitute of the caudal projection, commonly called "the Pope's nose," even the spine itself being deficient in the final vertebræ—a fact discovered by dissection. Their gait is very remarkable; they carry themselves so erect that they look as though they would fall backwards. They are very lively and active; the little pullets looking when they run about, as Mr. Tegetmeier says, like animated cricket balls. The saddle hackle is very abundant, and falls prettily at the back.

In size and weight, also length and colour of legs, the Rumpless Bantams vary, according to the variety which they most resemble in other points; but they should not weigh more than 16 to 20 ounces each, and as they are very compact, they look quite small. In colour, the most common resemble the Black-red Game Bantams, and have rather long, clean, willow legs. Several good specimens of these are kept in the Isle of Man. Another most attractive kind is Mottled; the hens black-and-white, the cocks the same, but having coloured saddles and hackles. These, also, are clean legged and single combed, and have won many first prizes for Mrs. Ricketts. There are also some Nankin coloured ones, like those first produced, still in existence, which are dark cinnamon all over, except that they have black flights, and blue legs. A very charming variety, and the latest produced, is the Silver-laced, which Mr. Garnet showed in 1889. This pretty little hen was perfectly marked, and like a Sebright in every respect, excepting lack of tail. The all black, and pure white varieties, too, with their coral red faces, single combs, short legs and long boots, and upright Penguin-like carriage, are very quaint and pleasing. The Black Booted Rumpless should have dark legs and beaks; the White ones, white legs and beaks.

Rumpless Bantams are as hardy as most breeds of Bantams. They are good layers, though very many of their eggs are unfertile; they rarely want to sit, but when they do, are careful mothers, and the chickens are easily reared. They are generally a long-lived race, and one little fellow, which took a prize at the last Dairy Show, is about ten years old, and as healthy and lively as ever; they also easily become so tame as to eat out of one's hand.

The Rumpless Bantams are such interesting little birds, that we should like to see them take a more prominent place in our exhibitions, but suppose that they are unpopular because a handsome and gracefully carried tail lends a charm, which these little "Manx fowls," as they are sometimes called, are without.

Probably, before very long, Spanish Bantams will be added to the already rather long list, as Mr. James C. Lyell has been engaged on their production about twelve years. The time and space that he can devote to poultry is very limited; thus he has only been able to raise six or eight chickens each year, and has had little choice of stock.

Unfortunately, after four or five seasons' breeding, the best bird, a cockerel, was worried by a dog, and Mr. Lyell had almost to begin again; but his patient perseverance has been so far rewarded that he now has a cock and four hens very little larger than Black Rosecomb Bantams, and with white faces, except above the eyes, where, formerly, many of the pure large Spanish failed. The combs of the hens are not yet quite satisfactory, being erect instead of folding over, and probably it will be some time before the ear lobes are as large in proportion as those of the large variety.

Briefly stated, the Spanish Bantam cock should have large, erect, deeply serrated comb; long, thin, pendant, red wattles; large, pure white face and ear lobe, which should not show any break or division, and should be as soft and free from wrinkles as a piece of white kid; the beak should be dark horn colour; the eyes wide open, bright, and dark; and the legs and feet as nearly black as possible. The whole of the plumage should be glossy raven black, and the carriage proud and majestic. The hen should be as nearly as possible like the cock, except that her comb should fold over like a Minorca's.

Last year, there was an attempt made to produce Houdan Bantams, and a very promising brood of first cross chickens had advanced half way to maturity, when, from some cause or other, they came to an untimely end.

We have also heard of Silver-grey Dorkings and White Creve-Cœur Bantams, but as neither of these kinds has been exhibited recently, we fear they have died out through lack of appreciation; for it needs a vast amount of perseverance to work steadily on at a new breed, with but scant encouragement from the "fancy" generally.

CHAPTER XVI.

STANDARDS FOR JUDGING BANTAMS.

WE will now give, in the same order in which they have already been written upon, a minute description of an ideally perfect Bantam of each variety.

Our readers must bear in mind that the correct weight for Bantams is one-fifth that of the large variety, and it is our endeavour, as nearly as possible, to give this weight, which they certainly should not exceed, though they may sometimes be slightly less; but there is always a danger of losing stamina, symmetry, and other essential points, by trying still further to reduce size.

GAME BANTAMS.

GENERAL CHARACTERISTICS.—COCK.

Head.—Long, lean, and narrow.
Comb.—(If undubbed) thin, small, neat, well serrated, perfectly erect, and straight.
Beak.—Long, strong, and slightly curved.
Eyes.—Large, bold, and prominent.
Face.—Lean, covered with a thin, fine skin.
Throat.—Long, with fine, smooth skin.
Wattles.—(If undubbed) small, round, and thin.
Ear Lobes.—(If undubbed) very small, and quite free from any white.
Neck.—Long and slender, forming, with the head, a graceful curve.
Hackle.—Short, close-fitting, and narrow.
Back.—Flat and short, broad at the shoulders, and tapering off to a fine, narrow stern.
Breast.—Rather broad, but not too prominent.
Wings.—Short, well curved, and fitting closely to the body.
Tail.—Small, carried tightly together, and rather low.
Sickle Feathers.—Fine and narrow.
Tail Coverts.—Narrow, short, and fine.
Thighs.—Long, muscular, set well apart, and rather forward.
Legs.—Long, straight, round, and slender, with fine, close-fitting scales.
Feet.—Sound, the toes long, straight, well spread out, flat on the ground, the hind toes set on low.
Plumage.—Short, hard, and bright.
Body in Hand.—Firm, muscular, and compact.
Weight.—As a cockerel, not more than 20 ounces; nor 24 ounces as a cock.
General Shape and Carriage.—Erect, tall, smart, and racy.

GENERAL CHARACTERISTICS.—HEN.

Head.—Long, narrow, tapering, and neat.
Beak.—Long, strong, and slightly curved.
Eyes.—Large, prominent, and brilliant.
Face.—Lean, covered with smooth, fine skin.
Comb.—Very small, thin, neat, perfectly erect and straight, low at front, and well serrated.
Wattles.—Small, fine, thin, and round.
Ear Lobes.—Very small and close, and quite free from any white.
Throat.—Long, with fine, smooth skin.
Neck.—Long, slender, and slightly arched.
Hackle.—Short, close fitting, and narrow.
Back.—Flat and short, broad at the shoulders, and tapering towards the tail.
Breast.—Rather broad, but not too prominent.
Wings.—Short, well curved, fitting closely to the body, and well up at the flights.
Tail.—Short and narrow, and carried rather low.
Thighs.—Long, firm, muscular, set well apart, and rather forward.
Legs.—Long, straight, slender, round, with fine scales.
Feet.— Placed flat on the ground, the toes long, thin, straight, and well spread, the hind toe well set back.
Plumage.—Short, hard, and bright.
Weight.—Pullets not more than 18 ounces, hens 20 ounces.

BROWN-BREASTED RED GAME BANTAMS.

Bred by, and the Property of, Mr. Dan Clayton, Box Tree Mill, Bradford.

Cock, winner of First Prizes at Crystal Palace, Birmingham, etc., etc., 1892.
Hen, First and Challenge Cup at Crystal Palace, First at Dairy Show, Birmingham, etc., etc.

General Shape and Carriage.—Smart, tall, upright, and fearless.

POINTS OF COLOUR IN BLACK-BREASTED RED GAME BANTAMS.—COCK.

Head.—Light orange-red.
Face and Throat.—Bright red.
Comb, Wattles, and Ear Lobes.—(If undubbed) Bright red.
Eyes.—Ruby red, *i.e.*, clear, rich, full red.
Beak.—Dark horn colour.
Neck Hackle.—Light orange-red, free from stripes.
Back.—Bright crimson.
Wing Butts and Shoulder.—Black.
Wing-bow and Shoulder Coverts.—Crimson.
Bars.—Steel blue.
Primaries.—Black, except the two lower feathers, the outer part of each being edged with bay.
Secondaries.—Clear bay outer web, and black inner web.
Saddle Hackle.—Orange, to match the neck.
Tail.—Black.
Sickle Feathers and Tail Coverts.—Glossy, blue-black.
Breast and Thighs.—Bluish-black.
Underpart of Body.—Black, and free from any rusty colour.
Legs and Feet.—Willow, or olive.

HEN.

Head.—Golden.
Face, Comb, Wattles, and Ear Lobes.—Red.
Eyes.—Full, rich, ruby red.
Beak.—Dark horn colour.
Neck Hackle.—Golden, with narrow, but well defined, black stripes.
Back, and Shoulder Coverts, Wing-bow, and Coverts.—All one even shade cf light brownish drab, finely pencilled with black.
Primaries.—Black.
Secondaries.—Outer edge like the wing-bow, the rest black.
Tail.—Black, except the upper feathers, which should be the same as the back and wings.
Throat.—Light salmon colour.
Breast.—Rich salmon red, shading off to ashy grey on the belly.
Thighs.—Pale salmon, or ashy grey.
Legs and Feet.—Willow, or olive.

POINTS OF COLOUR IN BROWN-BREASTED RED GAME BANTAMS.—COCK.

Head.—Lemon, striped with black.
Comb.—(If not dubbed) Dark purple.
Face.—Dark gipsy, *i.e.*, dark mulberry or purple.
Wattles and Ear Lobes.—(If undubbed) Dark red or purplish.
Eyes.—Black.
Beak.—The darkest horn colour.
Neck Hackle.—Lemon-coloured, with a very narrow black stripe down the centre of each feather.
Back, Shoulder Coverts, and Saddle.—Lemon.
Wing Butts and Shoulder.—Black.
Wing-bow.—Lemon.
Coverts.—Rich black.
Primaries and Secondaries.—Black.
Tail.—Black.
Sickle Feathers and Tail Coverts.—Glossy green-black.
Breast.—Rich black, each feather showing pale lemon-coloured shaft, and margin all round.
Thighs and Underparts.—The front may be laced like the breast, the rest must be black.
Legs, Feet, and Toe Nails.—The darkest bronze, almost black.

HEN.

Head.—Lemon, or straw colour.
Face, Comb, Wattles, and Ear Lobes.—Gipsy, *i.e.*, dark purple.
Eyes.—Black.
Beak.—Black, or the darkest horn.
Neck Hackle.—Straw colour, with narrow black stripes.
Breast.—Black, laced with straw colour.
Remainder of Plumage.—Lustrous green-black.
Legs and Feet.—Dark bronze, or black.

POINTS OF COLOUR IN BIRCHEN GREY GAME BANTAMS.—COCK.

Head.—Silvery white.
Comb, Face, Throat, Wattles, and Ear Lobes.—(If not dubbed) Dark purple, or mulberry.
Eyes.—Black, or as dark as possible.
Beak.—The darkest horn colour.
Neck Hackle.—Silvery white, with a very narrow black stripe down the centre of each feather.
Back, Shoulder Coverts, and Saddle.—Silvery white.
Wing Butts and Shoulder.—Black.
Wing-bow.—Silvery white.
Wing-bar or Coverts.—Glossy rich black.

Breast.—Rich black, each feather showing silvery white shaft, and margin all round, giving a beautifully laced appearance.

Thighs.—The front may be laced like the breast, the rest must be black.

Sickle Feathers and Tail Coverts.—Glossy green black.

Remainder of Plumage.—Sound black.

Legs, Feet, and Toe Nails.—The darkest bronze, almost black.

HEN.

Head, and Neck Hackle.—Silvery white, with narrow black stripes.

Face, Comb, Wattles, and Ear Lobes.—Dark purple, or mulberry.

Eyes.—Black, or as dark as possible.

Beak.—The darkest horn colour.

Breast.—Black, laced with silvery white.

Remainder of Plumage.—Lustrous green-black.

Legs, Feet, and Toe Nails.—The darkest bronze, almost black.

POINTS OF COLOUR IN GOLDEN DUCKWING GAME BANTAMS.—COCK.

Head.—Creamy white.

Comb, Face, and Throat.—Red.

Eyes.—Bright ruby red.

Beak.—Dark horn colour.

Neck Hackle.—Creamy white, quite free from any black stripes.

Back.—Clear and even orange, or golden yellow.

Wing Butts and Shoulder.—Black.

Wing-bow and Shoulder Coverts.—Clear orange, or golden yellow.

Wing-bar.—Steel blue.

Primaries.—Black, except the lower edge of the last feathers, which should be light brown.

Secondaries.—The lower web, clear white, with a blue-black spot on the end of each feather; the upper web, black.

Saddle Hackles.—Straw colour.

Breast and Thighs.—Sound blue-black.

Underpart of Body.—Sound black, free from grey.

Tail.—Blue-black.

Sickle Feathers and Tail Coverts.—Glossy blue-black.

Legs and Feet.—Willow, or olive-green.

HEN.

Head.—Silvery white.

Face, Comb, Wattles, and Ear Lobes.—Red.

Eyes.—Ruby red.

Beak.—Dark horn colour.

Neck.—White, with narrow black stripes.

Throat.—Light salmon colour.

Breast.—Rich salmon colour, shading off to ashy grey on the thighs and belly.

Back, Shoulder Coverts, Wing-bow, and Coverts.—French-grey, or very pale slate-coloured ground, very finely pencilled all over with black.

Primaries.—Black.

Secondaries.—The outer web, like the wing-bow; the inner one, black.

Tail.—Black, except the upper feathers, which should be like the wings and back.

Legs and Feet.—Willow, or olive-green.

POINTS OF COLOUR IN SILVER DUCKWING GAME BANTAMS.—COCK.

Head.—Clear silvery white.

Comb, Face, and Throat.—Red.

Eyes.—Bright red.

Beak.—Dark horn colour.

Neck Hackle.—Silvery white.

Back, Wing-bow, Shoulder Coverts, and Saddle Hackle. All silvery white.

Wing Butts and Shoulder.—Black.

Wing-bar.—Steel blue.

Primaries.—Black, except the lower edge of the last feathers, which should be almost white.

Secondaries.—The lower web, white; the upper, black.

Breast and Thighs.—Sound blue-black.

Underpart of Body, and Tail.—Sound black.

Sickle Feathers, and Tail Coverts.—Glossy blue-black.

Legs and Feet.—Willow, or olive-green.

HEN.

Head.—White.

Face, Comb, Wattles, and Ear Lobes.—Red.

Eyes.—Ruby red.

Beak.—Dark horn colour.

Neck Hackle.—White, with narrow black stripes.

Throat and Breast.—Pale salmon.

Thighs and Underpart.—Pale ashy grey.

Back, Shoulder Coverts, Wing-bow, and Coverts.—Light French-grey ground colour, very finely pencilled over with black.

Primaries.—Black.

Secondaries.—The outer web, like the wing-bow; the inner one, black.

Tail.—Black, except the upper feathers, which are like the wings and back.
Legs and Feet.—Willow, or olive green.

POINTS OF COLOUR IN RED PILE GAME BANTAMS.—COCK.

Head.—Orange-red.
Comb, Face, and Throat.—Red.
Eyes.—Bright red.
Beak.—Either yellow or horn colour. It must correspond with the legs and feet.
Neck Hackle.—Orange-red.
Back, Wing-bow, and Shoulder Covert.—Rich crimson.
Saddle Hackle.—Orange, or orange-red.
Wing Butts and Shoulders.—White.
Wing Coverts.—White.
Wing Primaries.—White.
Secondaries.—The outer web, chestnut or bay; the inner, white.
Breast and Thighs.—Pure milky white.
Tail.—White.
Legs and Feet.—Yellow is preferable, but willow may be allowed.

HEN.

Head.—Light golden.
Comb, Face, Ear Lobes, and Wattles.—Red.
Eyes.—Bright ruby red.
Beak.—Yellow, or horn colour.
Neck Hackle.—Golden, with narrow white stripes.
Breast.—Rich salmon red.
Remainder of Plumage.—As clear creamy white as possible.
Legs and Feet.—Yellow is preferred to willow.

POINTS OF COLOUR IN LEMON PILE GAME BANTAMS.—COCK.

Head, and Neck Hackle.—Pale lemon.
Face, Throat, and Comb.—(If undubbed) Red.
Eyes.—Bright red.
Beak, Legs, and Feet.—Yellow preferred.
Back, and Wing-bow.—Orange, or lemon.
Saddle Hackle.—Pale lemon.
Wing Secondaries.—Pale chestnut on outer web.
Remainder of Plumage.—Pure milk-white.

HEN.

Head, and Neck Hackle.—Pale lemon, with fine white stripes.
Comb, Face, Ear Lobes, and Wattles.—Red.
Eyes.—Bright red.

Beak, Legs, and Feet.—Yellow preferred.
Breast.—Pale salmon colour.
Remainder of Plumage.—Pure milk-white.

POINTS OF COLOUR IN WHITE GAME BANTAMS.

Comb, Face, Ear Lobes, and Wattles.—Red.
Eyes.—Rich ruby red.
Beak, Legs, and Feet.—Yellow.
Plumage.—Pure white.

POINTS OF COLOUR IN BLACK GAME BANTAMS.

Comb, Face, Ear Lobes, and Wattles.—Purplish red.
Eyes.—Dark brown, or black.
Beak.—Dark horn colour, or black.
Legs and Feet.—Dark bronze, almost black.
Plumage.—Lustrous black.

POINTS OF COLOUR IN RED WHEATEN GAME BANTAM.—HEN.

Comb, Face, Ear Lobes, and Wattles.—Red.
Eyes.—Rich red.
Beak.—Greenish horn.
Head, and Neck Hackle.—Golden.
Breast.—Pale fawn, or cream colour.
Thighs, and Underpart of Body.—Creamy, or light buff.
Back and Wings.—Pale cinnamon, or wheat colour.
Primaries.—Black.
Secondaries.—The outer web, wheaten; the inner, black.
Tail.—Black, the top outer feathers edged with wheaten colour.
Legs and Feet.—Willow, or olive green.

POINTS OF COLOUR IN DUCKWING WHEATEN GAME BANTAM.—HEN.

Comb, Face, Ear Lobes, and Wattles.—Red.
Eyes.—Rich red.
Beak.—Greenish horn.
Head, and Neck Hackle.—White, or white slightly striped with black.
Breast.—Light fawn.
Thighs, and Underpart of Body.—Creamy, or light buff.
Back and Wings.—Pale cinnamon.
Primaries.—Black.
Secondaries.—The outer web, wheaten; the inner, black.

Tail.—Black, the outer feathers edged with wheaten colour.
Legs and Feet.—Willow, or olive green.

POINTS OF COLOUR IN PILE WHEATEN GAME BANTAM.—HEN.

Comb, Face, Ear Lobes, and Wattles.—Red.
Eyes.—Rich ruby red.
Beak.—Horn colour, or yellow.
Head, and Neck Hackle.—Golden.
Breast.—Pale fawn, or cream colour.
Thighs, and Underpart of Body.—Creamy, or light buff.
Back and Wings.—Pale cinnamon, or wheat colour.
Primaries.—White.
Secondaries.—The outer web, wheaten; the inner, white.
Tail.—White, the outer feathers edged with wheaten colour.
Legs and Feet.—Yellow is preferable to willow.

SCALE OF POINTS FOR GAME BANTAMS.

A perfect bird to count 100 points, thus:—

	1892.	1872.
Good head and neck	10	10
Good eyes	8	4
Good body and wings	7	8
Good legs and feet	12	6
Good tail	10	8
Good symmetry	10	12
Good condition	10	18
Good colour of feather	15	20
Hardness and quality of feather	8	(hackle) 4
Correct size and weight	10	10
	100	100

POINTS TO BE DEDUCTED FOR DEFECTS.

Bad head and neck	10
Bad eyes	8
Bad body and wings	7
Bad legs and feet	12
Bad tail	10
Want of symmetry	10
Want of condition	10
Bad colour of feather	15
Too much feather	8
Incorrect size and weight	10
	100

DISQUALIFICATIONS.

Duck feet, crooked breasts, wry tail, and deformed back. If shown together in a pen, birds not matching in colour, etc.

MALAY BANTAMS.

GENERAL CHARACTERISTICS.—COCK.

Head.—Broad across the skull, with overhanging eyebrows.
Comb.—Strawberry, or half walnut, small and neat, and set well forward.
Beak.—Strong and hooked.
Eyes.—Bright, and deep set, having a cruel expression.
Face.—Skinny and bare.
Throat.—Bare skin running a long way down the neck.
Wattles and Ear Lobes.—Very small.
Neck.—Long, slightly curved, and carried very upright.
Hackle.—Very short and scanty, except at base of skull.
Back.—Long, sloping, and curved convexly.
Shoulders.—Very broad and prominent, carried well up, and generally bare of feathers at the points.
Wings.—Short, strong, well curved, clipping the body tightly.
Saddle.—Narrow and drooping, the feathers short and scanty.
Tail.—Of moderate length, drooping, with very narrow wiry sickles, slightly curved.
Breast.—Hard, deep and full, frequently bare of feathers on the point of the breast bone.
Thighs.—Long and muscular, scantily feathered, leaving the hocks quite bare.
Legs.—Long, massive and round, except at the hocks, where they should be flat; and covered with fine, close-fitting scales.
Feet.—The toes long and straight, the hind toes set on low, and the feet firmly placed flat on the ground.
Plumage.—Very short, hard and brilliant.
Body in Hand.—Extremely firm and muscular.
Weight.—As cockerels, not more than 28 ounces; nor 32 ounces as cocks.
General Shape and Carriage.—Tall, gaunt and fierce; high at shoulders; narrow and drooping at stern.

GENERAL CHARACTERISTICS.—HEN.

Head.—Broad across the skull, with massive eyebrows.
Comb.—Small, half walnut.
Beak.—Strong and hooked.
Eyes.—Fierce and deep set.
Face.—Skinny and bare.

Wattles and Ear Lobes.—Small and fine.
Throat.—Lean and red, slightly covered with small hair-like feathers.
Neck.—Long, slightly curved, and very upright.
Hackle.—Very short and scanty, except at base of skull.
Back.—Long, sloping, and rather convex in outline.
Shoulders.—Very broad, high and prominent.
Wings.—Short, strong, and well curved.
Tail.—Rather short and square, carried slightly above the horizontal line, and very flexible; constantly moving from side to side.
Breast.—Deep and full, the point often bare.
Thighs.—Long, muscular and scantily feathered.
Legs.—Long, massive and round; finely scaled.
Feet.—The toes long and straight, the hind toes set on low, and the feet firmly placed flat on the ground.
Plumage.—Short, hard and brilliant.
Body in Hand.—Very hard, firm and muscular.
Weight.—As pullets, not more than 22 ounces; nor 24 ounces as hens.
General Shape and Carriage.—Tall, gaunt, fierce, high at the shoulders, narrow and drooping at stern.

POINTS OF COLOUR IN BRIGHT RED MALAY BANTAMS.—COCK.

Comb, Face, Throat, Ear Lobes, and Wattles.—Brilliant red.
Eyes.—White, pearl, yellow, or daw.
Beak.—Yellow, or horn; yellow preferred.
Head, and Neck Hackle.—Orange red.
Back, Wing-bow, and Shoulder Coverts.—Rich crimson.
Wing Butts.—Black.
Bars.—Brilliant black.
Primaries.—Black, the lower web edged with bay.
Secondaries.—Clear bay outer web, and black inner web.
Saddle Hackle.—Orange red.
Tail.—Greenish-black.
Breast, Thighs, and Underparts.—Glossy black.
Legs and Feet.—Rich yellow.

HEN.

Comb, Face, Ear Lobes, and Wattles.—Red.
Eyes.—White, pearl, yellow, or daw.
Beak.—Yellow, or horn; yellow preferred.
Head, and Neck Hackle.—Rich golden, with narrow black stripes.
Back, Wings, and Tail Coverts.—Rich brown, finely pencilled all over with black.
Wing Primaries, and Tail.—Black.
Breast.—Rich salmon red, or cinnamon.
Thighs and Underparts.—Brown.
Legs and Feet.—Rich yellow.

POINTS OF COLOUR IN DARK RED MALAY BANTAM.—COCK.

Comb, Face, Throat, Ear Lobes, and Wattles.—Bright red.
Eyes.—White, pearl, yellow, or daw.
Beak.—Yellow preferred.
Head, and Neck Hackle.—Deep red.
Back, and Wing-bow.—Dark purplish crimson, or maroon.
Saddle Hackle.—Deep red.
Remainder of Plumage.—Very lustrous raven black.
Legs and Feet.—Rich yellow.

POINTS OF COLOUR IN CINNAMON MALAY BANTAM.—HEN.

Comb, Face, Ear Lobes, and Wattles.—Bright red.
Eyes.—White, pearl, yellow, or daw.
Beak.—Yellow preferred.
Head, and Neck Hackle.—Dark bay, or purplish.
Primaries.—Dark bay.
Secondaries.—The outer web, wheaten; the inner web, black, or dark bay.
Tail.—Dark bay; the upper outer feathers, cinnamon.
Remainder of Plumage.—Wheaten, or cinnamon, as even in shade as possible all through.
Legs and Feet.—Rich yellow.

POINTS OF COLOUR IN PHEASANT MALAY BANTAMS.—COCK.

Comb, Face, Throat, Ear Lobes, and Wattles.—Brilliant red.
Eyes.—White, pearl, yellow, or daw.
Beak.—Horn, or yellow striped with horn.
Head, and Neck Hackle.—Raven black.
Back, Saddle, and Saddle Hackle.—A mixture of rich, glossy green-black and dark maroon, black predominating.
Wing-bow.—Dark maroon.
Primaries.—Black.
Secondaries.—The outer web, rich bay; the inner web, black.
Tail.—Glossy green-black.

Remainder of Plumage.—Raven black, exceedingly lustrous.

Legs and Feet.—Rich yellow.

HEN.

Comb, Face, Ear Lobes, and Wattles.—Bright red.

Eyes.—White, pearl, yellow, or daw.

Beak.—Horn, or yellow striped with horn.

Head, and Neck Hackle.—Raven black.

Breast, Wings, Back, Thighs, and Tail Coverts.—One uniform rich bay, every feather distinctly laced with glossy, metallic green-black. Some are double laced, *i.e.*, one lacing inside the other, and others single laced; and either is correct; the lacing to be as regularly even as possible.

Tail.—Black.

Legs and Feet.—Rich yellow.

POINTS OF COLOUR IN PILE MALAY BANTAMS.—COCK AND HEN.

Comb, Face, Throat, Ear Lobes, and Wattles.—Brilliant red.

Eyes.—White, or pearl.

Beak and Legs.—Rich yellow.

Plumage.—Exactly the same on all parts as the Pile Game Bantams.

POINTS OF COLOUR IN WHITE MALAY BANTAMS.—COCK AND HEN.

Comb, Face, Throat, Ear Lobes, and Wattles.—Rich red.

Eyes.—White, or pearl.

Beak and Legs.—Rich yellow.

Plumage.—Pure milk-white, quite free from any straw, or other coloured feathers.

SCALE OF POINTS FOR MALAY BANTAMS.

A perfect bird to count 100 points.

POINTS TO BE DEDUCTED FOR DEFECTS.

Bad head and neck	15
Bad eyes	10
Want of shoulder	10
Bad legs and feet	10
Bad tail	5
Want of reach and symmetry	12
Want of condition	10
Defective colour	8
Too much, or long and soft feather	10
Incorrect size and weight	10
	100

DISQUALIFICATIONS.

Bad feet, triple combs, deformed back, wry tail, willow legs, crooked breasts, etc.

INDIAN GAME BANTAMS.

GENERAL CHARACTERISTICS.—COCK.

Comb.—Triple or pea comb, small, neat, and closely set on the head.

Head.—Rather long and thick, but neither as thick, nor the eyebrows as deep and prominent as the Malay's.

Beak.—Strong, well curved, and stout where it joins the head.

Eyes.—Full and bold.

Face.—Smooth, and fine in texture.

Throat.—Not as lean as English Game Bantams, and dotted with small feathers.

Wattles and Ear Lobes.—Very small, smooth, and neat.

Neck.—Nicely arched, and of medium length.

Hackle.—Short, but full, and not reaching below the base of the neck.

Back.—Rather short and flat, broad at the shoulders, and tapering towards the tail.

Breast.—Broad, moderately deep and prominent, and well rounded.

Wings.—Short and close, well rounded, rather high at the shoulders, and the ends neatly tucked up.

Tail.—Medium in length, and drooping; a few narrow, short, side sickles, and tail coverts; and the feathers close, hard, and glossy.

Thighs.—Stout, and well rounded, but shorter than the Malay's.

Legs.—Very strong and thick, quite upright from feet to hocks, and covered with fine, closely-fitting scales.

Feet.—Strong; the toes, long, straight, and well set apart; the hind toe, low; and the feet placed firmly on the ground.

Plumage.—Short, hard, and close.

Body in Hand.—Firm and muscular.

Weight.—Not more than 28 ounces as cockerels; nor 32 ounces as cocks.

General Shape and Carriage.—Broad and compact, but elegant; active, sprightly, vigorous and fearless.

GENERAL CHARACTERISTICS.—HEN.

Head.—Rather long and thick.

Comb.—Triple, or pea comb, small and neat, and as evenly serrated as possible.

Beak.—Strong, stout and well curved.

Eyes.—Full and bold.

Face and Throat.—Smooth and fine in texture, thinly covered with small hair-like feathers.

Wattles and Ear Lobes.—Very small, smooth and neat.
Neck.—Moderately long, and slightly arched.
Hackle.—Short, but rather full.
Back.—Rather short and flat, broad at the shoulders, and tapering towards the stern.
Breast.—Broad, rather deep and well rounded.
Thighs.—Round and muscular.
Legs.—Strong and thick, and covered with fine scales.
Feet.—The toes, long, straight, well set apart; the hind toe, low, and nearly flat on the ground.
Plumage.—Short, hard, and close.
Body in Hand.—Firm.
Weight.—As a pullet, not more than 22 ounces; nor 24 ounces as a hen.
General Shape and Carriage.—Rather broad and compact, yet elegant; bold, vigorous, active and sprightly.

POINTS OF COLOUR IN LACED INDIAN GAME BANTAMS.—COCK AND HEN.

Comb, Face, Ear Lobes, and Wattles.—Brilliant red.
Eyes.—White, or daw.
Beak.—Horn colour, or yellow striped with horn.
Legs.—Rich orange-yellow.
Plumage.—Exactly the same as that of the Pheasant Malay Bantams.

POINTS OF COLOUR IN WHITE INDIAN GAME BANTAMS.

Comb, Face, Ear Lobes, and Wattles.—Rich red.
Eyes.—White, or daw.
Beak and Legs.—Rich yellow.
Plumage.—Pure milk-white.

SCALE OF POINTS FOR INDIAN GAME BANTAMS.

A perfect bird to count 100 points.

POINTS TO BE DEDUCTED FOR DEFECTS.

Bad head and neck	15
Bad eyes	5
Badly shaped body and shoulders	15
Bad legs and feet	10
Bad tail	5
Want of symmetry	12
Want of condition	8
Defective colour	10
Want of hardness and brilliancy of feather.	10
Incorrect size and weight	10
	100

DISQUALIFICATIONS.

Single comb, legs any other colour than yellow, roach back, crooked breast, or any other deformity.

ASEEL BANTAMS.

GENERAL CHARACTERISTICS.—COCK.

Comb.—Triple, or pea comb, very small, hard and horny.
Head.—Rather small, but short and thick, and broad between the eyes and beak.
Beak.—Very strong, thick, and somewhat short; the lower mandible, straight; the upper one, slightly curved.
Eyes.—Bold, prominent and brilliant, and set well back in the head.
Face.—Fine and smooth, but hard skin.
Throat.—Smooth and lean, not prominent.
Ear Lobes and Wattles.—As small as possible.
Neck.—Hard, muscular and powerful, round, of medium length, the same width throughout, slightly curved at the back, and well seated between the shoulders.
Back.—Broad at the shoulders, short and quite straight.
Wings.—Short, strong, and carried quite level, standing well out from the shoulders, and showing a bare spot at the first joint.
Breast.—Short, broad and flat, almost naked at the point of the bone, and quite free from fluff.
Stern.—Narrow compared with the shoulders, but strong and thick at the root of tail.
Tail.—Short and slightly drooping; the feathers narrow and hard; the sickles very fine, short and nicely curved.
Tail Coverts.—Short, spare, wiry and very hard.
Thighs.—Strong, thick, muscular, set well apart, and sparingly feathered.
Legs and Feet.—Short, straight, clean, closely and evenly covered with scales; the toes and nails the same. A straight hind toe is preferable, but a "duckfoot" is not a disqualification.
Plumage.—Short, hard, wiry, and without any appearance of fluff.
Body in Hand.—Firm, hard, and evenly balanced.
Weight.—From 20 to 24 ounces.
General Shape and Carriage.—Straight, upright, angular, quick and agile.

GENERAL CHARACTERISTICS.—HEN.

The hen should resemble the cock as closely as possible; her ear lobes and wattles being scarcely perceptible, and the tail close and compact.

POINTS OF COLOUR.—COCK AND HEN.

Comb, Face, Ear Lobes, Wattles, and Throat.—Red.
Eyes.—Pearl, of all shades.
Beak and Legs.—Colour immaterial, provided they match.
Plumage.—Pure white, glossy raven black, black-red, grey, red spangle, black spangle, yellow, and pheasant, etc. They may be of any colour, as in this variety, shape and carriage are of more consequence; but if shown in pairs, they must always match in colour and marking.

SCALE OF POINTS FOR ASEEL BANTAMS.

A perfect bird to count 100 points.

POINTS TO BE DEDUCTED FOR DEFECTS.

Softness in condition	20
Bad carriage	20
Roach back	10
Coarse head and comb	10
Weak neck	10
Narrow, long stern	5
Upright tail	10
Bad plumage	5
Scars and blemishes	10
	100

DISQUALIFICATIONS.

Long thin body, long legs, rough scaly legs, level carriage, and squirrel tail.

CHAPTER XVII.

STANDARDS FOR JUDGING BANTAMS (CONTINUED). TECHNICAL TERMS.

PEKIN OR COCHIN BANTAMS.

GENERAL CHARACTERISTICS.—COCK.

Head.—Small, fine and neat.

Comb.—Single, fine, neat, and as small as possible; perfectly straight and erect, well serrated, and nicely curved from front to back.

Beak.—Rather short but stout, and slightly curved.

Eyes.—Bright and large.

Face.—Quite smooth, and fine in texture.

Wattles.—Long, ample, smooth and fine in texture, and neatly rounded.

Ear Lobes.—Smooth, fine, and well developed; nearly as long as the wattles.

Neck.—Short, full, nicely arched, and carried rather forward.

Hackle.—Very abundant and rather long, reaching well on to the back.

Back.—Short and broad, increasing in breadth to the *Saddle*, which should be very full, rise well from between the shoulders, and furnished with long, soft feathers.

Wings.—Small, very short, and tightly tucked up, the ends hidden by the saddle hackle.

Tail.—Very short, soft and full, having no hard quill feathers.

Tail Coverts.—Very abundant and nicely curved, almost hiding the tail; the whole tail forming one unbroken curve with the back and saddle.

Breast.—Very broad, deep, full, and rounded in appearance.

Thighs.—Short, broad, and set well apart.

Hocks.—Completely covered with soft feathers, which curl round the joint, and stand well out. Stiff feathers, called "vulture hocks," are objectionable, but not a disqualification.

Legs.—Short and thick, and abundantly covered with soft feathers, which stand well out.

Feet and Toes.—Strong and straight; the middle and outer toes covered with soft feathers quite to the ends, and as plentifully as possible.

Plumage.—Very abundant, long, and quite soft; the *Fluff*, which grows between the saddle and thighs, so full as to hide the latter.

Weight.—From 32 to 36 ounces.

General Shape and Carriage.—Broad, deep, plump, and well rounded; the carriage bold, rather forward but low, the head being not much higher than the tail.

GENERAL CHARACTERISTICS.—HEN.

Head.—Very small and neat.

Comb.—Small, neat, single, well serrated, perfectly erect and straight.

Beak.—Rather short and stout, slightly curved.

Eyes.—Bright, with a gentle expression.

Face.—Smooth, and fine in texture.

Wattles.—Fine skinned, small, round and thin.

Ear Lobes.—Smooth, fine, and well developed, almost as long as the wattles.

Neck.—Short, and carried rather forward.

Hackle.—Full and long.

Back.—Broad and short, rising into a full, round *Cushion*.

Tail.—Very small and soft, and almost hidden by the cushion.

Wings.—Very short, and well tucked up to the sides.

Breast.—Broad, deep, full, and rounded in appearance.

Thighs.—Short, broad, set wide apart, and well feathered.

Hocks.—Abundantly covered with soft feathers, having a tendency to curl inwards.

Legs.—Very short, and well feathered.

Feet and Toes.—Strong and straight; the middle and outer toes abundantly feathered quite to the ends. A short downy feathering on the inside of the shanks and inner toes is to be encouraged.

Plumage.—Very plentiful, long, and quite soft; the *Fluff*, very full, almost hiding the thighs; the toe and shank feathering as abundant as possible.

Weight.—From 28 to 32 ounces.

General Shape and Carriage.—Broad, deep, plump, and well rounded; the head carried slightly forward, and the breast very low, almost touching the ground, while the cushion is rather high.

POINTS OF COLOUR IN BUFF PEKIN BANTAMS.—COCK.

Comb, Face, Ear Lobes, and Wattles.—Bright red.

Eyes.—Red, or as nearly red as possible.

Beak.—Rich yellow.

Head, and Neck Hackle.—Rich, bright golden buff, perfectly even and sound in shade.

Back, Wing-bow, and Saddle Hackle.—The same shade as the neck hackle.

Tail.—Rich golden buff, quite free from any black, white, or bronze feathers.

Remainder of Plumage.—One even shade of rich golden buff, quite sound to the root of the feathers, and perfectly free from either black or white.

Legs and Feet.—Bright yellow.

HEN.

Comb, Face, Ear Lobes, and Wattles.—Bright red.

Eyes.—Red, or as nearly red as possible.

Beak and Legs.—Rich yellow.

Plumage.—Rich sound buff, perfectly even in shade throughout, and free from any black or white feathers.

Buff Pekins, both cocks and hens, are of three recognized shades, viz., light buff (or lemon buff), buff, and cinnamon. All are permissible, but whichever shade the bird is of, it should be as clear and even as possible all through.

POINTS OF COLOUR IN BLACK PEKIN BANTAMS.—COCK AND HEN.

Comb, Face, Ear Lobes, and Wattles.—Bright red.

Eyes.—Red, or yellow; red preferred.

Beak.—Dark horn colour, or black edged with yellow.

Legs and Feet.—Yellow preferred; but dark legs, with the underpart of the feet and toes yellow, are allowable.

Plumage.—One uniform, lustrous, beetle-green shade of black, sound down to the skin, and quite free from any white or coloured feathers.

POINTS OF COLOUR IN CUCKOO PEKIN BANTAMS.—COCK AND HEN.

Comb, Face, Ear Lobes, and Wattles.—Bright red.

Eyes.—Red.

Beak.—Orange-yellow, or yellow slightly marked with horn colour.

Plumage.—Light French-grey ground, every feather evenly and distinctly barred across several times with dark slate colour. The marking must be as fine and regular as possible, and the birds free from any white or straw-coloured feathers.

Legs and Feet.—Sound orange-yellow.

POINTS OF COLOUR IN PARTRIDGE COCHIN BANTAMS.—COCK.

Comb, Face, Ear Lobes, and Wattles.—Bright red.

Eyes.—Red, or orange; red preferred.

Beak.—Yellow, shading to horn colour.

Head.—Dark orange red.

Neck Hackle.—Bright orange, or golden red, becoming lighter in shade towards the shoulders; each feather distinctly striped down the centre with black.

Back, Shoulder Coverts, and Wing-bow.—Full rich crimson.

Wing Butts.—Black.

Greater and Lesser Coverts.—Lustrous green-black.

Primaries and Secondaries.—Bay on the outer web, and black on the inner.

Saddle Hackle.—Bright orange, each feather having a black stripe down the centre.

Tail.—Glossy green-black.

Breast, Thighs, and Underparts.—Sound black, with as much as possible of the beetle-green shading.

Foot and Shank Feather, and Fluff.—Good sound black; quite free from any rusty, grey, or white feathers.

Legs and Feet.—Rich yellow.

HEN.

Comb, Face, Ear Lobes, and Wattles.—Bright red.

Eyes.—Red, or orange; red preferred.

Beak.—Yellow, shading to horn colour.

CUCKOO & BLACK PEKIN BANTAMS.

Bred by, and the Property of, Mrs. Entwisle, Calder Grove House, near Wakefield.

Cuckoo Cockerel, winner of First and Special at Whitby, First and Special at Hallam and Ecclesall, etc., etc., 1894.

Black Hen, winner of First and Special at Whitby, First and Special at Driffield, etc., etc., 1892-3;
Firsts at Ripon, Middlesbro', etc., etc., 1894.

Head, and Neck Hackle.—Light golden, or straw colour; each feather distinctly striped with black.

Remainder of Plumage.—Clear, light, golden brown ground, finely and evenly pencilled all over with concentric rings of a dark shade—glossy green-black being preferred; the whole to be as uniform in shade and marking as possible.

Legs and Feet.—Rich yellow.

POINTS OF COLOUR IN WHITE COCHIN BANTAMS.—COCK AND HEN.

Comb, Face, Ear Lobes, and Wattles.—Bright red.

Eyes.—Red, or orange; red preferred.

Beak.—Rich yellow.

Plumage.—Pure snowy white, quite free from any creamy or yellow tinge, or splashes of black.

Legs and Feet.—Clear, rich yellow.

SCALE OF POINTS FOR PEKIN OR COCHIN BANTAMS.

A perfect bird to count 100 points.

POINTS TO BE DEDUCTED FOR DEFECTS.

Bad head and comb	10
Want of fluff and cushion	15
Too great length of leg	10
Deficiency of leg and foot feather	10
Defective colour	20
Bad shape and carriage	15
Want of condition	10
Incorrect size and weight	10
	100

DISQUALIFICATIONS.

Twisted or drooping comb; legs, any other colour than yellow (with the exception of the Blacks); slip wings, crooked backs, legs, or any other deformity; and when shown in pairs, birds not matching in colour, etc.; plucked tails, or other fraudulent trimming.

BRAHMA BANTAMS.

GENERAL CHARACTERISTICS.—COCK.

Comb.—Triple or pea, as small as possible, erect and firmly set; the centre ridge slightly the highest, but all three ridges perfectly straight and evenly serrated.

Head.—Small and well rounded, of medium breadth, rather short, and with slightly prominent eyebrows.

Beak.—Short, strong, and curved.

Eyes.—Large, and rather deep set.

Face.—Smooth; and gentle and docile in expression.

Wattles.—Smooth, well rounded, small, fine in texture, and free from hairs or feathers.

Ear Lobes.—Rather long in proportion to the wattles, quite smooth and fine, and free from feathers.

Neck.—Long, well arched, and showing a slight depression between the head and upper hackle feathers.

Hackle.—Ample, and flowing well on to the shoulders.

Back.—Broad, short and flat; the *Saddle* rising gradually from the middle of the back to the tail coverts.

Wings.—Of medium size, carried horizontally, and the ends well tucked up.

Tail.—Medium in length, full and well spread, and carried rather high.

Tail Coverts.—Abundant, and well curved.

Breast.—Very full, broad, deep, and carried well forward.

Thighs.—Large, powerful, and well feathered.

Hocks.—Abundantly covered with soft curling feathers, or with stiff feathers called "vulture hocks," provided the foot feathering corresponds; but the former is preferred.

Legs.—Of medium length, strong and wide apart; the scales as smooth and regular as possible, and the feathering very abundant.

Feet and Toes.—Large, straight, and well spread, heavily feathered to the ends of the middle and outer toes.

Plumage.—Very abundant, long, and soft; the *Fluff* standing well out from the thighs, and covering the hinder parts.

Weight.—As cockerels, not more than 2 lbs.; nor 2 lbs. 6 oz. as cocks.

General Shape and Carriage.—Broad, compact, deep, and square; the carriage, lofty, bold, and commanding.

GENERAL CHARACTERISTICS.—HEN.

Comb.—Exactly the same as the cock's, but smaller and closer fitting.

Head.—Small, well rounded, of medium breadth, rather short, and with slightly prominent eyebrows.

Beak.—Short, strong, and curved.

Eyes.—Large, and rather deep set.

Face.—Smooth ; and gentle in expression.

Wattles.—Smooth, well rounded, fine in texture, and free from feathers.

Ear Lobes.—Rather long in proportion to the wattles, quite smooth, fine, and free from feathers.

Neck.—Short, well arched, and showing a slight depression between the head and upper hackle feathers.

Hackle.—Ample, and flowing well on to the shoulders.

Back.—Broad, short, and flat; the *Cushion* rising gradually from the middle of the back to the tail coverts.

Wings.—Of medium size, carried horizontally, and the ends well tucked up.

Tail.—Medium in length, full, and well spread, and carried almost upright.

Tail Coverts.—Full, and almost covering the tail.

Breast.—Very full, broad, deep, and carried well forward.

Thighs.—Large, powerful, and heavily feathered.

Hocks. — Profusely covered with soft curling feathers, or with stiff feathers, called "vulture hocks," provided the foot feathering corresponds; but the former is preferred.

Legs.—Rather short, strong, and wide apart; the scales as smooth and regular as possible; and the feathering very abundant.

Feet and Toes.—Large, straight, and well spread, heavily feathered to the ends of the middle and outer toes.

Plumage.—Very abundant, long, and soft; the *Fluff* standing well out from the thighs, and covering the hinder parts.

Weight.—As pullets, not more than 28 ounces, nor 2 lbs. as hens.

General Shape and Carriage.—Broad, compact, deep, and square; the carriage sedate and gentle, yet erect and commanding.

POINTS OF COLOUR IN LIGHT BRAHMA BANTAMS.—COCK.

Comb, Face, Ear Lobes, and Wattles.—Bright coral-red.

Head.—Pure silvery white.

Beak.—Bright orange-yellow, or light horn colour.

Eyes.—Either red or yellow; red preferred.

Neck Hackle.—Silvery white, striped with sound black, the lower parts being more densely striped than the upper.

Saddle Hackle.—Silvery white; slightly, but evenly, striped with black.

Primaries.—Black, or black edged with white.

Secondaries.—White on the outer web, black on the inner web.

Tail.—Black.

Tail Coverts.—Glossy black; slightly, but evenly, edged with white.

Leg and Foot Feathers.—As white as possible on the surface, but with distinct black in the underparts of the feather.

Remainder of Plumage.—White; if possible, white down to the skin; but frequently the fluff, or down, is dark, and this is not objected to, provided it does not show through the feather.

Legs and Feet.—Orange-yellow.

HEN.

Comb, Face, Ear Lobes, and Wattles.—Bright coral-red.

Head.—Pure silvery white.

Beak.—Yellow, or light horn colour.

Eyes.—Red, or yellow; red preferred.

Neck Hackle.—Silvery white, striped with sound black, the stripes being bolder on the lower part, where each feather must have a black centre, entirely surrounded by a white margin.

Tail.—Black.

Tail Coverts.—Black, with a white edging round each feather.

Primaries.—Black, or black edged with white.

Secondaries.—White on the outer web, and black on the inner web.

Leg and Foot Feathers.—Chiefly white, but what black is visible should be distinct.

Remainder of Plumage.—Pure silvery white, quite free from black ticks or markings on the surface. If the fluff, or down, near the skin is dark, it is not objected to, provided it does not show through the feather; but white down is preferred.

Legs and Feet.—Orange-yellow.

POINTS OF COLOUR IN DARK BRAHMA BANTAMS.—COCK.

Comb, Face, Ear Lobes, and Wattles.—Bright coral-red.
Head.—Silvery white.
Eyes.—Red, or yellow; red preferred.
Beak.—Dark horn colour.
Neck Hackle.—Silvery white, distinctly striped with brilliant black; the stripes increasing in breadth down to the shoulders and back.
Back, and Shoulder Coverts.—Silvery white, except between the shoulders, where the feathers should be black, laced with white.
Wing Butts and Shoulders.—Brilliant black.
Wing-bow.—Pure silvery white.
Wing Coverts.—Rich beetle-green black.
Primaries.—Black.
Secondaries.—White on the outer web, except a black spot on the ends; and black on the inner web.
Saddle Hackle.—Silvery white, well striped with brilliant black, the stripes increasing in breadth on to the tail coverts.
Tail.—Black.
Tail Coverts.—Glossy beetle-green black, the two upper ones edged with white.
Breast, Thighs, and Underparts.—Brilliant black.
Leg and Foot Feathers.—As black as possible.
Legs and Feet.—Orange-yellow.

HEN.

Comb, Face, Ear Lobes, and Wattles.—Bright coral-red.
Head.—Silvery white.
Eyes.—Red, or yellow; red preferred.
Beak.—Dark horn colour.
Neck Hackle.—Silvery white, very clearly and evenly striped with rich black; a bold stripe to be preferred, and the marking to come well up to the throat, and level with the under side of the eye at the back of the neck.
Tail.—Black, the top outer feathers slightly edged with grey.
Remainder of Plumage.—Either light French-grey ground, pencilled with steel grey; or, steel grey, pencilled with rich black. The ground colour must be of one shade throughout, and the pencilling in concentric rings, following the outline of the feather, as distinct and uniform as possible all over, including the shank and toe feather.
Legs and Feet.—Orange-yellow.

SCALE OF POINTS FOR BRAHMA BANTAMS.

A perfect bird to count 100 points.

POINTS TO BE DEDUCTED FOR DEFECTS.

Bad head and comb	10
Want of fluff and cushion	15
Deficiency of leg and foot feather	10
Bad shape and carriage	15
Defective colour	20
Want of symmetry	8
Want of condition	12
Incorrect size and weight	10
	100

DISQUALIFICATIONS.

Comb, other than triple; legs, any colour but yellow; twisted hackle, crooked back, or other deformity; plucked tails, or any other fraudulent trimming; and when shown in pairs, birds not matching fairly well.

BOOTED BANTAMS.

GENERAL CHARACTERISTICS.—COCK.

Comb.—Single, of medium size, well serrated, and perfectly erect and straight.
Head.—Small and neat.
Face.—Smooth, fine skin, free from hairs.
Eyes.—Large, bright, and prominent.
Beak.—Rather stout, and of medium length.
Ear Lobes.—Small and neat.
Wattles.—Small, round, and close.
Neck.—Rather short, and curved; the head carried well back.
Breast.—Full and prominent.
Wings.—Large, long, and carried drooping.
Back.—Short.
Tail.—Large, long, and full; carried well forward over the back, and abundantly furnished with sickles.
Tail Coverts.—Ample, and nicely curved.
Thighs.—Short, and well feathered.
Legs.—Moderately short, and heavily feathered with long and rather stiff feathers, those growing from the hocks almost touching the ground.
Feet and Toes.—Well spread, and straight; very heavily feathered on the outer and middle toes.
Plumage.—Long and abundant.
Weight.—From 20 to 26 ounces.
Carriage.—Erect, and strutting.

GENERAL CHARACTERISTICS.—HEN.

Comb.—Single, small, well serrated, and perfectly erect and straight.
Head.—Small and neat.
Face.—Smooth, fine skin, free from hairs.
Eyes.—Large, bright, and prominent.
Beak.—Rather stout, and of medium length.
Ear Lobes.—Small and neat.
Wattles.—Very small, and neatly rounded.
Neck.—Rather short, and curved, the head carried well back.
Breast.—Full and round.
Wings.—Large, long, and carried drooping.
Back.—Short.
Tail.—Large, full, well spread, and carried slightly over the back.
Thighs.—Short, and well feathered.
Legs.—Moderately short, and heavily feathered with long, rather stiff feathers, those growing from the hocks almost touching the ground.
Feet and Toes.—Well spread and straight, very heavily feathered on the outer and middle toes.
Plumage.—Long and abundant.
Weight.—Not less than 16 ounces, nor more than 22 ounces.

POINTS OF COLOUR IN WHITE BOOTED BANTAMS.—COCK AND HEN.

Comb, Face, Ear Lobes, and Wattles.—Bright red.
Eyes.—Red.
Beak and Legs.—White.
Plumage.—Pure white.

POINTS OF COLOUR IN BLACK BOOTED BANTAMS.—COCK AND HEN.

Comb, Face, Ear Lobes, and Wattles.—Bright red.
Beak.—Black, or dark horn.
Eyes.—Dark red, or very dark brown.
Legs and Feet.—Black.
Plumage.—Black, as rich and lustrous as possible.

POINTS OF COLOUR IN SPANGLED BOOTED BANTAMS.—COCK AND HEN.

Comb, Face, Ear Lobes, and Wattles.—Bright red.
Beak.—Horn colour.
Eyes.—Dark red, or brown.
Legs and Feet.—Dark.
Plumage.—Black and white; red and white; or black, white, and bright red; as evenly distributed as possible over the whole body.

SCALE OF POINTS FOR BOOTED BANTAMS.

A perfect bird to count 100 points.

POINTS TO BE DEDUCTED FOR DEFECTS.

Bad head, comb, and face	15
Defective colour of plumage	20
Wrong colour of legs and beak	10
Deficiency of leg and foot feather	15
Want of symmetry	15
Want of condition	10
Incorrect size and weight	15
	100

DISQUALIFICATIONS.

Any deformity, fraudulent trimming, and comb other than single.

SULTAN BANTAMS.

GENERAL CHARACTERISTICS. COCK AND HEN.

Crest.—Large, compact, and globular.
Comb.—Very small, consisting of two minute spikes standing up near the beak, and almost hidden by the crest.
Beak.—Rather short, and curved.
Eyes.—Bright and lively.
Muffling.—Very full and compact, the whiskers and beard being joined in one.
Ear Lobes and Wattles.—Very small and neat, the former hidden by muffling.
Neck.—Rather short, well arched, and heavily feathered; carried slightly back.
Back.—Short and straight.
Wings.—Rather short, and drooping.
Tail.—The cock's tail should be full, with long flowing sickles; the hen's, rather large and expanded; both should be carried high.
Breast.—Full, deep, and prominent.
Thighs.—Very short, and furnished with heavy vulture hocks, covering the joint.
Legs and Feet.—Short, and well covered with abundant soft feather to the end of the toes, which should be five in number, and quite straight.
Plumage.—Long, soft, and very abundant.
General Shape and Carriage.—Deep, but neat and compact; the carriage, brisk and sprightly.
Weight.—Cocks, from 16 to 20 ounces; hens, 12 to 14 ounces.

POINTS OF COLOUR IN SULTAN BANTAMS.
COCK AND HEN.

Comb and Wattles.—Bright red.
Beak.—White.
Eyes.—Red.
Legs and Feet.—White, or pinky white.
Plumage.—Pure white throughout.

SCALE OF POINTS FOR SULTAN BANTAMS.

A perfect bird to count 100 points.

POINTS TO BE DEDUCTED FOR DEFECTS.

Deficiency in size, or faulty shape of crest	15
Want of beard and whiskers	10
Deficiency of leg and foot feathers	15
Legs wrong colour, or too long	10
Defective colour of plumage	20
Want of symmetry	12
Want of condition	8
Incorrect size and weight	10
	100

DISQUALIFICATIONS.

Coloured feathers in plumage; wry tail, crooked back, or other deformity; and absence of foot feather.

BURMESE BANTAMS.

GENERAL CHARACTERISTICS.
COCK AND HEN.

Crest.—Full, but falling over the back of the head.
Comb.—Small, single, well serrated, straight, and erect, placed well forward, and in front of the crest.
Eyes.—Bright and prominent.
Beak.—Short and strong.
Ear Lobes.—Very small and neat.
Wattles.—Rather long and pendant in the cock; as small as possible in the hen.
Neck.—Short and thick; the hackle very abundant and long.
Back.—Short and flat.
Wings.—Long and drooping, the ends touching the ground.
Tail.—The cock's, very large and long, profusely furnished with long, finely tapering sickles; the hen's, long and well spread, and both to be carried rather high.
Breast.—Broad, deep, and full.
Thighs, Legs, and Feet.—Extremely short, and heavily feathered, the toe feather on the outsides four or five inches in length.
Plumage.—Very abundant.
General Shape and Carriage.—Deep, broad, and long; the carriage lively.
Weight.—Cocks, about 20 ounces; hens, 16 ounces.

POINTS OF COLOUR IN BURMESE BANTAMS.

Comb, Face, Ear Lobes, and Wattles.—Bright red.
Beak.—Yellow, or horn.
Eyes.—Red, or orange.
Legs and Feet.—Rich yellow, excepting the blacks, which have black legs, with yellow under the feet and between the toes.
Plumage.—Pure white, rich black, brown, grey, or speckled.

SCALE OF POINTS FOR BURMESE BANTAMS.

A perfect bird to count 100 points.

POINTS TO BE DEDUCTED FOR DEFECTS.

Bad head and comb	10
Deficiency of crest	10
Scanty hackle	5
Too small or badly carried tail	15
Deficiency of foot feather	10
Faulty colour of plumage	20
Want of symmetry	12
Want of condition	8
Incorrect size and weight	10
	100

DISQUALIFICATIONS.

Crooked back, wry tail, or other deformity; and absence of crest or foot feather.

SILKY BANTAMS.

GENERAL CHARACTERISTICS.
COCK AND HEN.

Crest.—Of the cock, pointing backwards; of the hen, globular.
Comb.—Strawberry, or half walnut, small, and placed between the beak and crest.
Beak.—Short and curved.
Eyes.—Bright, but not very prominent.
Ear Lobes.—Of medium size.
Wattles.—Of the cock, rather long and pendant; the hen's being small and round.
Head.—Moderate in size.

Neck.—Medium in length; the hackle very abundant, and flowing well on to the shoulders.

Back.—Very short, the *Saddle* of the cock being very broad, and rising almost from the neck hackle, and falling again at the tail, the *Cushion* of the hen being well rounded.

Tail.—Very small, and short.

Tail Coverts.—Extremely soft and abundant.

Wings.—Short, and closely carried.

Breast.—Broad, full, and round.

Thighs.—Short, and well covered with *Fluff.*

Legs and Feet.—As short and as profusely feathered as possible, with vulture hocks; the toes, five in number, also being feathered, and well spread out.

Plumage.—Soft, silky hairs and down instead of feathers, as ample and fine in quality as possible.

General Shape and Carriage.—Compact, rather strutting, and active.

Weight.—The cocks should not be more than 15 ounces, nor the hens more than 12 ounces.

POINTS OF COLOUR IN SILKY BANTAMS. COCK AND HEN.

Comb, Face, Wattles, and Skin of Body.—Bluish purple.

Ear Lobes.—Sky blue.

Beak.—Bluish white.

Eyes.—Very dark, almost black.

Legs and Feet.—Black, or dark slate blue.

Plumage.—Either pure white, intense black, golden brown, or silvery grey; the colour to be quite sound and even throughout.

SCALE OF POINTS FOR SILKY BANTAMS.

A perfect bird to count 100 points.

POINTS TO BE DEDUCTED FOR DEFECTS.

Bad head and comb	10
Badly-shaped crest	5
Defective colour of skin	15
Bad ear lobes	10
Insufficient foot feather	10
Bad colour of plumage	15
Want of symmetry	15
Want of condition	10
Incorrect size and weight	10
	100

DISQUALIFICATIONS.

Deformity of any kind; absence of crest, or fifth toe; plumage, other than silky; and red or white skin.

SEBRIGHT BANTAMS.

GENERAL CHARACTERISTICS. COCK AND HEN.

Comb.—Rose, broad in front, and tapering into a long spike at the back; full of serrations, as even as possible; firmly set on the head, flat on the top, and pointing slightly upwards.

Head.—Small, and carried well back.

Eyes.—Bright and full.

Beak.—Rather short.

Face.—Smooth, fine skin.

Ear Lobes.—Quite smooth and soft, and of medium size.

Wattles.—Of medium size, and well rounded.

Neck.—Rather short and thick; carried well back.

Back.—Very short and flat.

Wings.—Rather large, and carried very low, and drooping.

Tail.—Rather large, carried high and open, so as to show every feather.

Breast.—Very prominent and full.

Thighs.—Short.

Legs.—Short, slender, perfectly free from feather, and covered with closely fitting scales.

Feet and Toes.—Straight, slender, and well spread.

Plumage.—Short, and well rounded at the ends; both sexes are feathered exactly alike, the cock having neither sickles nor saddle hackle.

General Shape and Carriage.—Compact, vain, jaunty, and strutting, yet nervous and tremulous.

Weight.—Cocks, about 22 ounces; hens, 18 ounces.

POINTS OF COLOUR IN GOLD SEBRIGHT BANTAMS.—COCK AND HEN.

Comb, Face, Ear Lobes, and Wattles.—Dark purple, or mulberry.

Eyes.—Black, or as dark as possible.

Beak.—Dark horn.

Legs and Feet.—Slate blue.

Plumage.—The ground colour clear almond, or golden bay, evenly laced all round with rich greenish black, every feather being distinctly edged with black.

POINTS OF COLOUR IN SILVER SEBRIGHT BANTAMS.—COCK AND HEN.

Comb, Face, Ear Lobes, and Wattles.—Dark purple.

Eyes.—Black, or as dark as possible.

Beak.—Dark horn colour.

Legs and Feet.—Slate blue.

GOLD SEBRIGHT BANTAMS.

The Property of Lady Alington, Crichel, Wimborne.

Cock, winner of First at Dairy Show, First and Cup at Crystal Palace, etc., etc., 1893.
Hen, First at Crystal Palace, etc., 1893.

Plumage.—Pure silvery white, every feather distinctly and evenly laced with rich greenish black.

POINTS OF COLOUR IN CREAMY SEBRIGHT BANTAMS.—COCK AND HEN.

Comb, Face, Ear Lobes, and Wattles.—Dark purple.
Eyes.—Black, or nearly so.
Beak.—Dark horn.
Legs and Feet.—Slate blue.
Plumage.—Creamy white, evenly laced throughout with rich greenish black.

SCALE OF POINTS FOR SEBRIGHT BANTAMS.

A perfect bird to count 100 points.

POINTS TO BE DEDUCTED FOR DEFECTS.

Bad comb	10
Bad face and ear lobes	12
Splashed tail	15
Defective lacing	25
Faulty colour	20
Want of symmetry	5
Want of condition	8
Incorrect size and weight	5
	100

DISQUALIFICATIONS.

Single comb, wry tail, or any other deformity; feathers on legs, sickle feathers in the cocks, or any feathers devoid of lacing.

ROSECOMB BANTAMS.

GENERAL CHARACTERISTICS.—COCK.

Comb.—Rose, broad at the front, and tapering into a long spike, slightly pointing upwards at the back, the surface level, and well and evenly serrated.
Head.—Short and broad.
Beak.—Short, and slightly curved.
Eyes.—Full.
Face.—Smooth.
Ear Lobes.—Flat, smooth, soft, and almost circular.
Wattles.—Round, and fine in quality.
Neck.—Short, thick, and slightly arched; the hackle very abundant, and flowing well towards the middle of the back.
Back.—Moderately long, and broad; the saddle hackle very profuse.
Tail.—Large, and well spread; carried upright, but with a slight backward tendency, and amply furnished with broad, long sickles.
Wings.—Large, and slightly drooping, but not as much so as in the Sebrights.
Breast.—Broad, full, and prominent.
Thighs.—Short.
Legs.—Rather short and slender.
Feet.—Small, the toes slender, straight, and well spread.
General Shape and Carriage.—Compact, yet lively, graceful, and jaunty.
Weight.—From 16 to 20 ounces.

GENERAL CHARACTERISTICS.—HEN.

Comb.—Rose, small and neat, broad at the front, and tapering into a long spike, with an upward tendency at the back; full of serrations, as even as possible.
Head.—Short and broad.
Beak.—Short, and slightly curved.
Eyes.—Full.
Face.—Smooth.
Ear Lobes.—Smooth, soft, round, and flat.
Wattles.—Thin, and well rounded.
Neck.—Short, and slightly arched.
Back.—Moderately long, and broad.
Tail.—Large, well spread, and carried rather high.
Wings.—Large, and slightly drooping.
Breast.—Broad, full, and prominent.
Thighs.—Short.
Legs.—Short and slender.
Feet.—Small, the toes slender, straight, and well spread.
General Shape and Carriage.—Compact, lively and graceful, yet nervous and tremulous.
Weight.—From 14 to 17 ounces.

POINTS OF COLOUR IN BLACK ROSECOMB BANTAMS.—COCK AND HEN.

Comb, Face, and Wattles.—Rich, bright red.
Ear Lobes.—Pure white.
Eyes.—Bright, dark red, or ruby red.
Beak.—Dark horn colour.
Legs and Feet.—Black, or almost black, when young, and dark slate colour after the second year.
Plumage.—Rich, greenish black, as lustrous as possible.

POINTS OF COLOUR IN WHITE ROSECOMB BANTAMS.—COCK AND HEN.

Comb, Face, and Wattles.—Rich, bright red.
Ear Lobes.—Pure white.
Eyes.—Bright red.
Beak.—White.
Legs and Feet.—White.
Plumage.—Pure white.

SCALE OF POINTS FOR ROSECOMB BANTAMS.

A perfect bird to count 100 points.

POINTS TO BE DEDUCTED FOR DEFECTS.

Bad comb	20
Defective ear lobes and face	20
Insufficient length of feather	10
Defective colour	15
Bad shape and carriage	10
Want of symmetry	5
Want of condition	10
Incorrect size and weight	10
	100

DISQUALIFICATIONS.

Single comb, red ear lobes, crooked tail or back, squirrel tail, and plumage other than intense black or milk white.

CUCKOO, OR SCOTCH GREY BANTAMS.

GENERAL CHARACTERISTICS. COCK AND HEN.

Comb.—Perfectly erect, and either single or rose; a well serrated, medium sized, single comb preferred.
Head.—Moderately long and narrow.
Face.—Smooth.
Eyes.—Full and bright.
Beak.—Short, and slightly curved.
Ear Lobes.—Smooth, round, and neat.
Wattles.—Medium in size, and well rounded.
Neck.—Moderately long; with full, flowing hackle, in the cock.
Back.—Rather short, straight, and broad between the shoulders.
Wings.—Of medium size, and carried well up.
Tail.—The cock's, large and full, well furnished with flowing, nicely arched sickles, and carried slightly backwards; the hen's, broad and full, and carried rather upright.
Breast.—Full and round.
Thighs.—Long, and set well apart.
Legs.—Medium in length, and slender.
Feet.—Of medium size, the toes well spread.
Plumage.—Rather long and broad.
Weight.—Of cocks, from 16 to 20 ounces; of hens, 14 to 17 ounces.
General Shape and Carriage.—Compact, erect, and sprightly.

POINTS OF COLOUR IN CUCKOO BANTAMS. COCK AND HEN.

Comb, Face, Ear Lobes, and Wattles.—Brilliant red.
Eyes.—Rich red.
Beak.—White, or horn colour; white preferred.
Plumage.—Pale French-grey ground colour, very finely marked with distinct bands, across each feather, of dark slate-grey, the colour and marking to be as clearly defined as possible. Black feathers in hackles or saddle are very objectionable.
Legs and Feet.—White, or white spotted with black.

SCALE OF POINTS FOR CUCKOO BANTAMS.

A perfect bird to count 100 points.

POINTS TO BE DEDUCTED FOR DEFECTS.

Bad head and comb	15
White in ear lobes	10
Defective colour and marking	25
Bad feet	10
Want of symmetry	15
Want of condition	15
Incorrect size and weight	10
	100

DISQUALIFICATIONS.

Any deformity, white ear lobes, legs other than white or mottled; red, white, or many black feathers in plumage, or any fraudulent trimming.

NANKIN BANTAMS.

GENERAL CHARACTERISTICS. COCK AND HEN.

Comb.—Either single or rose.
Head.—Small and neat.
Eyes.—Bright and full.
Beak.—Short and small.
Face.—Smooth, and fine in texture.
Ear Lobes.—Smooth, flat, and nearly circular.
Wattles.—Thin, and well rounded.
Neck.—Arched, and of medium length; the cock's hackle very long and full, flowing well on to the shoulders.

Back.—Moderate in length.
Wings.—Long, and carried drooping.
Tail.—Large, and carried upright; the cock's tail well furnished with flowing sickles and coverts.
Breast.—Full, prominent, and carried well forward.
Thighs.—Short, neat, and slender.
Legs.—Short and slender.
Feet.—Small, the toes straight and well spread.
Weight.—Cocks, 18 to 20 ounces; hens, 14 to 16 ounces.
General Shape and Carriage—Compact, sprightly, and strutting.

POINTS OF COLOUR IN NANKIN BANTAMS
COCK.

Comb, Face, Ear Lobes, and Wattles.—Bright red.
Eyes.—Rich red.
Beak.—Horn colour.
Legs and Feet.—Blue.
Neck Hackle, Back, and Shoulder Coverts, Wing-bow, and Saddle.—Rich, dark cinnamon.
Tail.—Bronze shading into dark brown, or black.
Wing Primaries.—Black.
Secondaries.—The outer web, cinnamon; the inner, black.
Remainder of Plumage.—Rich, warm buff or cinnamon, as even as possible in shade.

HEN.

Comb, Face, Ear Lobes, and Wattles.—Bright red.
Eyes.—Rich red.
Beak.—Horn colour.
Legs and Feet.—Blue.
Wing Primaries.—Dark brown or black.
Tail.—Black, the upper outer feathers cinnamon.
Remainder of Plumage.—Clear medium shade of buff, as even as possible throughout.

SCALE OF POINTS FOR NANKIN BANTAMS.

A perfect bird to count 100 points.

POINTS TO BE DEDUCTED FOR DEFECTS.

Bad head and comb	20
Defective colour of plumage	25
Bad colour of legs	10
Want of symmetry	15
Want of condition	15
Incorrect size and weight	15
	100

DISQUALIFICATIONS.

Deformity of any kind, and white feathers in wings or tail.

G.

JAPANESE BANTAMS.

GENERAL CHARACTERISTICS.
COCK AND HEN.

Comb.—Single, large, deeply serrated, and perfectly erect and straight.
Head.—Of medium size, carried erect.
Face.—Smooth and fine.
Eyes.—Large and bright.
Beak.—Short and stout.
Ear Lobes.—Of medium size, and neat.
Wattles.—Large, and well rounded.
Neck.—Short, and carried well back; the cock's hackles very abundant, and flowing over the shoulders.
Back.—Short and broad; the cock amply furnished with saddle hackle.
Wings.—Long, broad, and carried with the points almost touching the ground.
Tail.—Large, long, and carried perfectly erect; the cock's amply furnished with long, fine sickles and coverts.
Breast.—Full, well rounded, and prominent.
Thighs.—Exceedingly short, and wide apart.
Legs.—As short as possible.
Feet.—Well spread; the toes straight and slender.
Weight.—Cocks, 18 to 22 ounces; hens, 14 to 18 ounces.
General Shape and Carriage.—Compact, erect, and strutting; the head and tail almost touching each other, while the wing ends are very near the ground.

POINTS OF COLOUR IN JAPANESE BANTAMS.
COCK AND HEN.

Comb, Face, Ear Lobes, and Wattles.—Bright red.
Eyes.—Red.
Beak and Legs.—Rich orange-yellow.
Plumage.—Black-tailed White, Black, White, Speckled, Buff, Grey, Brown, and Cuckoo.

The Black-tailed Whites have pure white bodies, with black tails, the sickles and tail coverts beautifully and evenly edged all round with white; the wing primaries and secondaries having a black inner web. The Speckled variety should be black and white, distributed as evenly as possible. The Greys have a black ground colour; the cock's hackles, back, and wing-bow, silver; and the hen having a fine silver lacing all over, and rather heavier on the neck hackles. The Cuckoos should, in colour and marking, closely resemble the Scotch Greys; and the Blacks, Whites and Buffs should be perfectly sound and even throughout, and quite free from any other coloured feathers.

SCALE OF POINTS FOR JAPANESE BANTAMS.

A perfect bird to count 100 points.

POINTS TO BE DEDUCTED FOR DEFECTS.

Bad head and comb	10
Too great length of leg	15
Defective colour of plumage	20
Bad shape and carriage of body	15
Bad carriage of tail	10
Want of symmetry	12
Want of condition	8
Incorrect size and weight	10
	100

DISQUALIFICATIONS.

Deformity of any kind; comb other than single; and when shown together in one pen, birds not matching in colour, etc.

FRIZZLED BANTAMS.

GENERAL CHARACTERISTICS.
COCK AND HEN.

Comb.—Either single, or rose; single preferred.
Head.—Small and neat.
Face.—Smooth, and fine in texture.
Eyes.—Bright and full.
Beak.—Short and strong.
Ear Lobes.—Moderate in size.
Wattles.—Pendulous, and well shaped.
Neck.—Rather short, and nicely arched.
Back.—Broad and short.
Wings.—Rather long, and drooping.
Tail.—Large, full, and erect, the cock having plenty of side hangers, and a good pair of sickles.
Breast.—Full, and well rounded.
Thighs.—Moderately short, and set well apart.
Legs.—Very short, and quite free from feathers.
Feet.—Well spread; the toes four in number.
Plumage.—Very short, hard, and wiry, every feather being curled backward towards the head, and the curling as close and abundant as possible.
Weight.—Cocks, 16 to 18 ounces; hens, 14 to 16 ounces.
General Shape and Carriage.—Compact, erect, lively, and strutting.

POINTS OF COLOUR IN FRIZZLED BANTAMS.
COCK AND HEN.

Comb, Face, Ear Lobes, and Wattles.—Bright red.
Eyes.—Brilliant red.
Beak.—Yellow, or horn.
Legs and Feet.—Yellow for the Whites and the Golden; dark willow, or black, for the dark feathered varieties.
Plumage.—Pure white, sound black, rich golden, partridge, grey, and blue. Whichever colour the birds are of, they should be as even as possible throughout.

SCALE OF POINTS FOR FRIZZLED BANTAMS.

A perfect bird to count 100 points.

POINTS TO BE DEDUCTED FOR DEFECTS.

Bad head and comb	5
Bad feet	5
Defective colour of plumage	15
Insufficiency of curl	25
Feather too soft	20
Want of symmetry	10
Want of condition	10
Incorrect size and weight	10
	100

DISQUALIFICATIONS.

Deformity of any kind; many feathers uncurled; and if shown together in one pen, birds not matching in colour, comb, etc.

POLISH BANTAMS.

GENERAL CHARACTERISTICS.
COCK AND HEN.

Crest.—As large and full as possible, globular in form, rising upright from the beak, and crescent shape in outline. The hen's crest should be very compact, and round.
Head.—Of medium size, hidden by crest.
Eyes.—Bright and full.
Beak.—Moderately large.
Comb.—Two minute horns, completely hidden by crest.
Ear Lobes.—Small, neat, and almost circular in the White-crested Polish; entirely concealed by *Muffling* in the other varieties.
Wattles.—Rather long and pendulous in the White-crested Polish; completely hidden by *Beard* in all the other varieties.

Neck.—Moderate in length, upright, and the cock's well furnished with hackle.
Back.—Straight, and medium in length.
Wings.—Large, but neatly tucked up.
Tail.—Large and full; that of the cock carried nearly upright; the hen's, rather open, or fanned.
Breast.—Full and round.
Thighs.—Short.
Legs.—Medium in length, and slender.
Feet.—Well spread; the toes straight and slender.
Weight.—Cocks, from 17 to 22 ounces; hens, from 14 to 18 ounces.
General Shape and Carriage.—Slight, but compact; erect, strutting, and vain.

POINTS OF COLOUR IN GOLD POLISH BANTAMS.—COCK.

Crest.—Black at the roots of each feather, golden bay in the centre, and tipped with black at the ends, and as free from white as possible.
Beard and Whiskers.—Black, laced with golden bay during the first year; afterwards, golden bay, laced with black.
Hackle.—Rich bay, tipped with black.
Saddle.—Bright, rich dark bay, hatched with lustrous black.
Wing-bow.—Deep, rich bay, spotted, and hatched with black.
Wing Coverts.—Golden bay, each feather laced with black, and thus showing two distinct lines across the wing.
Primaries and Secondaries.—Golden bay, laced with black.
Breast, Thighs, Back, and Shoulder Coverts.—Golden bay, each feather distinctly laced, and spangled at the tips with black.
Tail.—Golden bay, laced with black, and heavily spangled at the tips.
Sickle Feathers.—Rich bronze, and golden bay centre, evenly laced with glossy black, and well spangled at the ends.
Tail Coverts.—Rich bronze and golden bay, deeply laced all round with lustrous black.
Legs and Feet.—Blue.

HEN.

Crest.—During the first year black, well laced with golden bay, and entirely free from any white feathers; afterwards, golden bay, laced with black.
Beard and Whiskers.—The same as the crest.
Neck.—Golden bay, each feather distinctly laced at the end with black.
Wing Coverts, Primaries, and Secondaries.—Golden bay, evenly laced with black.
Tail.—Golden bay, heavily laced with black, and the lacing broader at the points.
Remainder of Plumage.—Clear, golden bay, evenly and distinctly laced, or spangled with black.
Legs and Feet.—Blue.

POINTS OF COLOUR IN SILVER POLISH BANTAMS.—COCK AND HEN.

Plumage.—Pure white ground colour, laced, or spangled with black; the markings being exactly the same as those described for the Golden Polish.
Legs and Feet.—Blue.

POINTS OF COLOUR IN CHAMOIS POLISH BANTAMS.—COCK AND HEN.

Plumage.—Exactly like the Golds in marking, the ground colour being clear golden bay, laced, and tipped with white.
Legs and Feet.—Blue.

POINTS OF COLOUR IN WHITE-CRESTED BLACK POLISH BANTAMS.

COCK AND HEN.

Crest.—Pure white, as free from black feathers as possible.
Face and Wattles.—Bright red.
Ear Lobes.—Pure white.
Remainder of Plumage.—Sound black, as rich and lustrous as possible.
Legs and Feet.—Blue, or black.

POINTS OF COLOUR IN BLACK, WHITE, BUFF, CREAMY, CUCKOO, BLUE, AND WHITE-CRESTED BLUE POLISH BANTAMS.

The Plumage of all these varieties should be as pure, and even in colour, as possible, the *Legs and Feet* of all being blue.

SCALE OF POINTS FOR POLISH BANTAMS.

A perfect bird to count 100 points.

POINTS TO BE DEDUCTED FOR DEFECTS.

Want of shape or size in crest	20
Want of muffling	10
Too much comb	15
Defective colour and marking	25
Want of symmetry	10
Want of condition	10
Incorrect size and weight	10
	100

DISQUALIFICATIONS.

Deformity of any kind, absence of muffling (except in the White-crested Blacks), legs other than blue, and if shown together in one pen, birds not matching in colour, etc.

MINORCA BANTAMS.

GENERAL CHARACTERISTICS. COCK AND HEN.

Comb.—Large, well serrated, and single; perfectly erect in the cock, and coming well over the back of the head; and doubling over on one side in the hen.

Face.—Smooth, and fine in texture.

Head.—Small and neat.

Ear Lobes.—Large, and like an almond in shape.

Wattles.—Large, and pendant.

Eyes.—Bold, and sparkling.

Beak.—Strong, and slightly curved.

Neck.—Of medium length, strong, and thick.

Back.—Broad and straight, and in the hen very long.

Wings.—Large, carried horizontally, and neatly tucked up.

Tail.—Large, the sickles and tail coverts of the cock flowing, and well curved.

Breast.—Very full and broad.

Thighs.—Large and strong.

Legs.—Rather short and well rounded, covered with smooth, fine scales.

Feet.—Firmly placed on the ground, the toes straight, and well spread.

Weight.—Cocks, from 18 to 22 ounces; hens, from 16 to 20 ounces.

General Shape and Carriage.—Of cock, compact, but erect and graceful; the hen being large, and square in body.

POINTS OF COLOUR IN MINORCA BANTAMS. COCK AND HEN.

Comb, Face, and Wattles.—Bright red.

Ear Lobes.—Pure white.

Eyes.—Rich, dark red.

Beak.—Of the Black variety, dark horn colour; of the White breed, white.

Legs and Feet.—Black, or dark slate colour, for the Blacks; white, for the Whites.

Plumage.—Brilliant, beetle-green black, as rich and lustrous as possible, for one variety; and pure, snow white for the other.

SCALE OF POINTS FOR MINORCA BANTAMS.

A perfect bird to count 100 points.

POINTS TO BE DEDUCTED FOR DEFECTS.

Bad head and comb	30
Stained ear lobes	10
Defective colour	25
Want of symmetry	10
Want of condition	15
Incorrect size and weight	10
	100

DISQUALIFICATIONS.

Deformity of any kind, the cock's comb falling over, and legs any colour than dark slate, black (or white for the White birds).

ANDALUSIAN BANTAMS.

GENERAL CHARACTERISTICS. COCK AND HEN.

In shape and size, these so closely resemble the Minorca Bantams, that it is unnecessary to give a detailed description of them.

POINTS OF COLOUR IN ANDALUSIAN BANTAMS.—COCK AND HEN.

Comb, Face, and Wattles.—Bright red.

Ear Lobes.—Pure white.

Beak.—Dark horn colour.

Eyes.—Sparkling, and very dark.

Legs and Feet.—Dark slate colour.

PLUMAGE OF COCK.

Neck Hackle, Back, Wing-bow, and Saddle Hackle.—Dense velvety black, or soft blue-black; the latter preferred.

Tail.—Light slate colour, or bluish grey.

Sickle Feathers and Tail Coverts.—A darker shade of slate blue.

Remainder of Plumage.—Light slate colour, or bluish grey, distinctly laced with black, or dark purple.

PLUMAGE OF HEN.

Neck.—Black, or dark blue-black.

Remainder of Plumage.—Light blue-grey, quite clear, and even throughout, every feather distinctly laced with black, or dark purple.

SCALE OF POINTS FOR ANDALUSIAN BANTAMS.

A perfect bird to count 100 points.

POINTS TO BE DEDUCTED FOR DEFECTS.

Bad head and comb	20
Badly shaped or stained lobes	10
Light eye	10
Defective colour and lacing	30
Want of symmetry	10
Want of condition	10
Incorrect size and weight	10
	100

DISQUALIFICATIONS.

Deformity of any kind, drooping comb in the cocks, red ear lobes, legs other than dark slate colour, birds not matching in show pen, etc.

LEGHORN BANTAMS.

GENERAL CHARACTERISTICS. COCK AND HEN.

Comb.—Single, free from side excrescences, and deeply serrated; perfectly erect in the cock, and extending well over the beak and back of the head; doubling over on one side in the hen.

Head.—Short and deep.

Eyes.—Full and bright.

Beak.—Rather long and stout.

Face.—Smooth, and fine in skin.

Ear Lobes.—Large, oval shaped, and smooth.

Wattles.—Large, and pendant.

Neck.—Of medium length, thick, and in the cocks covered with abundant hackle.

Back.—Broad, long, and flat.

Wings.—Large, but carried well up.

Tail.—Large, and the cock well furnished with sickles.

Breast.—Very deep, and full.

Thighs.—Short, but strong.

Legs.—Medium in length.

Feet.—Moderately large, the toes straight, and well spread.

Weight.—Cocks, 18 to 22 ounces; hens, 16 to 20 ounces.

General Shape and Carriage.—Deep, and massive, and rather long in keel; proud, stately, but active.

POINTS OF COLOUR IN LEGHORN BANTAMS. COCK AND HEN.

Comb, Face, and Wattles.—Bright red.

Ear Lobes.—White, or creamy in tint.

Eyes.—Brilliant red.

Beak.—Rich yellow.

Legs and Feet.—Rich orange-yellow.

Plumage.—Brown (closely resembling the Black-red Game Bantams in colour), White, Black, Buff, Duckwing, Pile (the two latter like Duckwing and Pile Game Bantams), Cuckoo, and Tri-coloured, or Mottled.

SCALE OF POINTS FOR LEGHORN BANTAMS.

A perfect bird to count 100 points.

POINTS TO BE DEDUCTED FOR DEFECTS.

Bad head and comb	20
Badly shaped or stained lobes	15
Defective colour and marking	15
Pale legs	15
Light eyes	5
Want of symmetry	10
Want of condition	10
Incorrect size and weight	10
	100

DISQUALIFICATIONS.

Any deformity, legs other than yellow, twisted comb of cock, and, when shown together, birds not matching in colour, etc.

HAMBURGH BANTAMS.

GENERAL CHARACTERISTICS. COCK AND HEN.

As these should be the same in all points as the Black and White Rosecomb Bantams, it is unnecessary again to repeat them here.

POINTS OF COLOUR IN GOLD-PENCILLED HAMBURGH BANTAMS.—COCK.

Comb, Face, and Wattles.—Brilliant red.

Ear Lobes.—Pure white.

Beak.—Horn colour.

Eyes.—Bright red.

Head, and Neck Hackle, Back, Wing-bow, Saddle Hackle, Breast, and Thighs.—Rich gold, as even in shade as possible.

Wing Coverts.—Gold; the upper web, and the tip of the lower web, pencilled with black.

Underparts.—Gold, slightly pencilled with black.

Tail.—Black.

Sickle Feathers, and Tail Coverts.—Brilliant black, laced all round with golden bay.

Wing Primaries and Secondaries.—Gold on the outer web, and black on the inner web.

Legs and Feet.—Dark slate colour, or blue.

HEN.

Comb, Face, and Wattles.—Brilliant red.

Ear Lobes.—Pure white.

Eyes.—Bright red.

Beak.—Horn colour.

Neck.—Light golden bay.

Primaries and Secondaries.—Golden bay, edged with black.

Remaining Plumage.—Light golden bay ground colour, each feather regularly pencilled across with several black bars.

Legs and Feet.—Dark slate blue.

POINTS OF COLOUR IN SILVER-PENCILLED HAMBURGH BANTAMS.—COCK AND HEN.

Plumage.—Exactly the same as the Golds, a pure silvery white ground colour being substituted for the golden bay.

Comb, Ear Lobes, &c.—Also of the same colours as the Golden Pencilled variety.

POINTS OF COLOUR IN GOLDEN-SPANGLED HAMBURGH BANTAMS.—COCK.

Comb, Face, and Wattles.—Bright red.

Ear Lobes.—Pure white.

Eyes.—Brilliant red.

Beak.—Horn colour.

Head, and Neck Hackle.—Rich golden bay, distinctly striped with black.

Breast, Thighs, and Underparts.—Golden bay ground colour, with a round, rich black spangle on the end of each feather. The spangling should be as distinct as possible.

Back, and Shoulder Coverts.—Reddish bay, each feather having a narrow black spot.

Wing-bow.—Bay, spangled with black.

Wing Coverts.—Golden bay, with a rich black round spangle at the end of each feather, forming two distinct bars across the wing.

Primaries and Secondaries.—Bay, with a black spot on the end of each feather.

Saddle Hackles.—The same as the neck hackles.

Tail.—Rich black, the sickles and tail coverts very brilliant beetle-green black.

Legs and Feet.—Dark slate-blue.

HEN.

Comb, Face, and Wattles.—Bright red.

Ear Lobes.—Pure white.

Eyes.—Red.

Beak.—Horn colour.

Head and Neck.—Bay, striped with black.

Tail.—Black.

Tail Coverts.—Golden bay, each feather tipped with black.

Remaining Plumage.—Golden bay, each feather heavily spangled at the end with black, the spots being round and distinct.

Legs and Feet.—Dark slate-blue.

POINTS OF COLOUR IN SILVER-SPANGLED HAMBURGH BANTAMS.—COCK AND HEN.

These are the same as the Golden Spangled, excepting that the ground colour is silvery white, spangled with rich black, and the tail white, with a large black spangle on the end of each feather.

SCALE OF POINTS FOR HAMBURGH BANTAMS.

A perfect bird to count 100 points.

POINTS TO BE DEDUCTED FOR DEFECTS.

Bad head and comb	15
Stained ear lobes	10
Bad carriage of tail	10
Defective colour and marking	30
Want of symmetry	10
Want of condition	15
Incorrect size and weight	10
	100

DISQUALIFICATIONS.

Deformity of any kind, single comb, red ear lobes, or legs other than blue.

RUMPLESS BANTAMS.

The chief characteristic of these birds is their *entire absence of tail.* In all other respects, both in

RUMPLESS BANTAMS.

Bred by, and the Property of, Mrs. Ricketts, Knighton Vicarage, Radnorshire.

Cock, winner of Second Prize at Birmingham; and the Hen, V. H. C. at Crystal Palace, 1893.

shape and colour, they may vary so considerably, that it would be an unnecessarily long task to attempt to enumerate them all. Suffice it to say that the correct *Weight* is from 16 to 20 ounces.

General Shape.—Round and compact.

Carriage.—Very erect, lively, and active.

SPANISH BANTAMS.

GENERAL CHARACTERISTICS. COCK AND HEN.

Comb.—Large, deeply serrated, single, perfectly erect in the cock, and folding over on one side in the hen.

Wattles.—Long, thin, and pendant.

Face, and Ear Lobes.—Large (*i.e.*, broad and deep), smooth, and soft, quite free from any break or division, and wrinkles.

Beak.—Strong, and slightly curved.

Eyes.—Large, and very bright.

Neck.—Of medium length, and well arched.

Back.—Moderate in length, and straight.

Wings.—Large, and neatly carried.

Tail.—Large, and carried high; the sickles and tail coverts of the cock, long, flowing, and abundant.

Breast.—Full, and round.

Thighs.—Of medium length, and round.

Legs.—Moderately long, round, and slender.

Feet.—Rather large, the toes slender, and well spread.

General Shape and Carriage.—Square and compact, erect, majestic, and stately.

Weight.—Cocks, from 18 to 22 ounces; hens, 16 to 20 ounces.

POINTS OF COLOUR IN SPANISH BANTAMS.

Comb, and Wattles.—Bright red.

Face, and Ear Lobes.—Pure white, resembling a piece of white kid.

Eyes.—Dark.

Beak.—Dark horn colour.

Plumage.—Lustrous raven black.

Legs and Feet.—Black, or as dark as possible.

SCALE OF POINTS FOR SPANISH BANTAMS.

A perfect bird to count 100 points.

POINTS TO BE DEDUCTED FOR DEFECTS.

Bad comb	15
Too small, or stained face and lobes	35
Defective plumage	15
Want of symmetry	10
Want of condition	15
Incorrect size and weight	10
	100

DISQUALIFICATIONS.

Wry back, crooked tail, or any other deformity, coloured feathers, spotted face, etc.

TECHNICAL TERMS.

A brief explanation of the terms in common use in connection with Bantams, will, no doubt, be appreciated by some of our readers, and is accordingly given in alphabetical order.

Beard.—A cluster of feathers growing under the beak of some varieties, as in the Polish Bantams.

Breed.—Any distinct variety of Bantams.

Brood.—A family of chickens under the care of one hen.

Broody.—Applied to a hen when wanting to sit, or incubate.

Carriage.—The natural position of a bird when alert.

Chick.—A newly-hatched fowl.

Chicken.—This word may be applied to a Bantam of any age under six months.

Cockerel.—A male bird under twelve months old.

Comb.—The red fleshy protuberance on the top of the head. (No. 1.)

Condition.—The state of a bird's health, shewn by its beauty of plumage and lively bearing.

Crest.—A tuft of feathers on the head.

Crop.—The bag or receptacle for food, on the breast of a bird.

Cushion.—The mass of feathers growing on the hinder part of a hen's back, and covering the tail. It is the most fully developed in a Cochin Bantam.

REFERENCES.

1. COMB.
2. FACE.
3. WATTLES.
4. EAR LOBES.
5. HACKLE.
6. BACK.
7. SADDLE.
8. SADDLE HACKLE.
9. TAIL COVERTS.
10. SICKLES.
11. TAIL FEATHERS.
12. SHOULDER BUTTS.
13. SHOULDER COVERTS.
14. WING-BOW.
15. WING COVERTS, OR "BAR."
16. SECONDARIES.
17. PRIMARIES, OR FLIGHTS.
18. BREAST.
19. THIGHS.
20. HOCKS.
21. LEGS, OR SHANKS.
22. SPURS.
23. TOES.

Dubbed.—The comb, lobes, and wattles having been remove , thus leaving the head smooth and clean.

Duck-footed.—The term applied to a bird when the hind toe, instead of being straight out behind in a line with the middle toe, lies close to the inner side of the foot.

Ear Lobes.—The folds of skin hanging below the ear; generally either red or white. (No. 4.)

Face.—The bare skin extending from the beak to the ears, and round the eyes. (No. 2.)

Flights.—The longest and strongest feathers of the wing, used when flying, but concealed by the "secondaries" when at rest. (No. 17.)

Fluff.—The soft, downy feathers growing on the thighs, and beneath the stern.

Furnished.—Matured, and fully feathered.

Gipsy Face.—The skin of the face being of a dark purple, or mulberry colour.

Hackles.—The narrow feathers growing on the neck of both sexes (No. 5), and also on the saddle of a cock; but the latter are more correctly called "saddle hackles" (No. 8), the word "hackles" alone, applying to the neck feathers.

GOOD FOOT — BAD OR "DUCK FOOT"

Hock.—The knee, or elbow joint of the leg. (No. 20.)

Leg.—The shank, or scaly part. (No. 21.)

Leg Feathers.—The feathers growing on the shanks. (*Ex.*, Pekins, Booted, etc.)

Metallic.—The lustre on the plumage, which resembles that on polished metal

Mossy.—Indistinct, or confused marking on the feathers.

Pea-comb.—A triple comb, composed of three parallel ridges, the middle one being the highest: for example, Brahma, and Indian Game Bantams.

Pencilling.—Small markings on the feathers, which may be either in lines across them, or in crescentic form, and following the shape of the feathers.

Primaries.—The flight feathers of the wing. (No. 17.)

Pullet.—A hen not more than twelve months old.

Reach.—The height of a bird when standing quite erect.

Rust.—A patch of reddish-brown on the wings of some breeds of pullets. This is also called "foxey."

Saddle.—The lower part of the back. (No. 7.)

Saddle Hackles.—The long, narrow feathers on the cock's back, falling over the root of the tail and wing ends. (No. 8.)

Secondaries.—The quill feathers of the wing, which cover the primaries, and are visible when the bird is at rest. (No. 16.)

Serrated.—Indented. The upper part of a single comb may have from three to eight spikes, and the spaces between are called serrations.

Shaft.—The stem, or quill, of a feather.

Shank.—The lower part of the leg. (No. 21.)

Shoulder Butt.—The rounded end of the wing. (No. 12.)

Sickles.—The long, curved feathers of a cock's tail. Usually, this term is applied only to the top pair, but is sometimes used for one or two pairs below. (No. 10.)

Slip-winged.—The primaries projecting over the secondaries.

Spangling.—The marking produced by each feather having a spot on the end of it of a different colour from the ground colour.

Spur.—The sharp, horny growth just above the foot of a Bantam. (No. 22.)

Squirrel Tail.—One that projects over the back, in front of a perpendicular line.

Strain.—A race of birds which, having been carefully bred for years, has attained an individual character of its own, which can be more or less relied upon.

Stained Lobes.—The ear lobes marked or spotted with another colour.

Style.—The general shape and appearance.

Symmetry.—The perfect agreement of one part with another. This is quite distinct from *Carriage;* and thus a bird may be perfect in proportion, and yet carry himself awkwardly.

Tail Feathers.—The straight, stiff feathers of the tail. (No. 11.)

Tail Coverts.—The soft, glossy, curved feathers that surround and cover the roots of the tail. (No. 9.)

Thigh.—The first joint of the leg, covered with feathers, and above the shank. (No. 19.)

Trio.—A cock and two hens, or cockerel and two pullets.

Under-colour.—The colour of the plumage seen when the surface is lifted, chiefly the down at the root of each feather.

Vulture Hock.—The stiff, projecting feathers at the hock joint.

Wattles.—The red, pendant skin below the beak. (No. 3.)

Web.—Of the feather, the flat or plume part; of the foot, the thin skin connecting the toes; of the wing, the triangular piece of skin seen when the wing is extended.

Whip Tail.—The feathers lying very close together, and folding over each other.

Wing-bar.—Any band of colour across the middle of the wing, caused by the marking upon, or colour of, the feathers known as *wing coverts*. (No. 15.)

Wing-bow.—The upper or shoulder part of the wing. (No. 14.)

Wing Coverts.—Two rows of broad feathers covering the roots of the secondaries. (No. 15.)

Wry Tail.—One that inclines either to the right or left, and is not carried in a straight line with the head.

CHAPTER XVIII.

PREPARATION OF BANTAMS FOR SHOWING, SENDING TO THE SHOWS, AND TREATMENT ON THEIR RETURN FROM THE SHOWS.

THIS subject is of vital interest, for what true fancier is there, who, possessing some good specimens of his favourite breed of Bantams, does not aspire to be a successful exhibitor, at least once or twice in the year? Yet, this he cannot hope to be, in these days of keen competition, unless his birds are properly prepared.

The first thing is, of course, to select those birds nearest to the required standard in colour, shape and size; and if they are in perfect health, and have been carefully housed, and kept on a good grass run, very little preparation will be necessary; this, however, varies according to the variety. For instance, a Game Bantam cockerel (unless shown when very young at some of the summer shows in June and July), should always be dubbed, after which operation he will not be presentable for three or four weeks; while a Black Pekin cockerel, if sufficiently tame, would need only his face and comb, etc., washing.

The use of the training pen, mentioned in the first chapter of this work, is indispensable, and the birds should be accustomed to it gradually from an early age, by shutting them up in it for an hour or so in the morning, and feeding them there. They should, also, be petted and stroked, and taught to feel confidence in their attendant, which they will not be long in showing, so that they will even eat out of the hand when in their run. They should then, by the aid of a light cane, be taught to hold themselves in such positions as best display their reach and style. In training Game Bantams, dainty bits held between the bars of the cage, only just within reach, help to make them show their height; but they must be carefully watched, and not allowed to acquire the habit of walking on their toes when in a cage, as some do: their feet should always be placed flat on the ground; but provided this is the case, the bird cannot stand too erect.

When Bantams have to be shown in pairs (that is, cock and hen together), they should be caged together for several days previously, so as to become thoroughly accustomed to each other's society.

The feeding is another important point to consider. Game, Malay, Indian Game, and Aseel Bantams require to be fairly plump, but very firm, and hard in flesh. They should have a feed of wheat, or oats, night and morning, and a light mid-day feed of Liverine, prepared as previously directed, or "Old Calabar's" foods, which are very largely used at shows. For other varieties, where hardness of flesh is not so much a feature as a nice, plump appearance, if the birds have this already, they may be fed in the same way; but if thin, they should be fed more generously, and more frequently, until the desired condition is attained. In the event of their being too fat, it would be well to discontinue the soft food, giving only grain, and small seeds twice daily. A handful of linseed may be given twice a week, as it helps to make the plumage glossy. Fresh water (to which may be added a few drops of sulphate of iron), sharp grit, and a liberal supply of green-food should also be daily provided; and scrupulous cleanliness maintained, the bottom of the cage being covered with chaff, or sawdust, which can be frequently renewed.

Immediately before sending Bantams to an exhibition, their faces, combs, etc., and legs and feet should be washed. First sponge the head, using clean water, and afterwards carefully dry it with a soft towel. The legs and feet often need the application of a little soap, and sometimes the use of a soft nail brush, to remove all dirt from them. A little vaseline, or pure olive oil, gently rubbed over the comb, face, and wattles, improves their appearance; then with a light rub down with an old silk handkerchief to give the final gloss to their feathers, they may be placed in the hamper for their journey.

WASHING BANTAMS.

White Bantams, and, indeed, all light coloured ones, need to be washed all over before they are sent to an exhibition. We have seen attempts made to wash the soiled parts only, when the whole plumage did not look dirty; but this invariably left a blotchy appearance, and would not tempt a second trial. A badly-washed bird, naturally, looks far worse than one shown just as it is taken from its run; but by the exercise of a little forethought, and by carefully following our directions, nobody need fear failure.

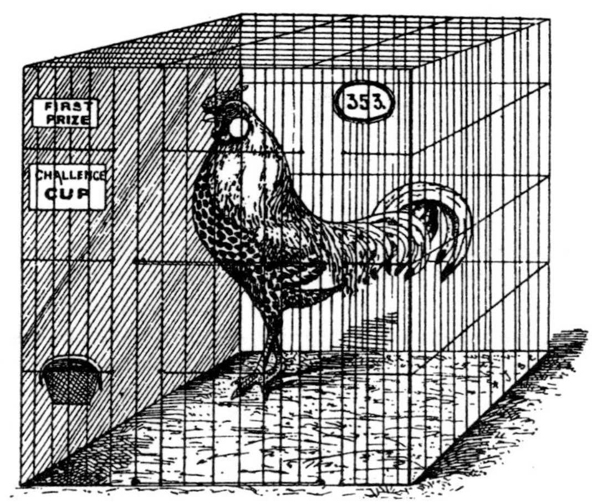

EXHIBITION OR TRAINING PEN.
(Wicks & Sons, Norwich.)

First see that you have a clear, bright fire, and abundant supply of hot, soft water. On no account use hard water, as it would cause the soap to curd, and completely spoil the bird. Partly fill two bowls, or tubs, with water as hot as one can comfortably bear one's hands in—one bowl for washing, the other for rinsing. We find round earthenware milk pans, about 8 inches deep and 18 to 20 inches wide, answer the purpose admirably.

The operator should provide himself with a large apron, covering him to the chest, and roll up his sleeves above the elbow; because, sometimes, the bird objects to the process, and gets one wing loose, a struggle and splash being the result. The bowls should be placed at a convenient height, then with a piece of white soap (we always use "Sunlight") and sponge, make a slight lather.

Before immersing the bird, always wash its head, throat, and the upper part of the neck thoroughly clean with the soapy sponge, but do not leave any lather on the face. Then take it in both hands, and hold it, up to the throat, under water for a few seconds, to saturate the feathers and prevent them breaking. Next hold the bird in the left hand, its head towards the left shoulder, its breast resting on the palm, and legs securely held between the fingers, soap the sponge, and thoroughly rub all over the neck, back, wings, under the wings, and breast. It will not hurt the short feathers to be rubbed sideways, or round and round; but the flights and tail must be rubbed always in one direction. By this time the bird will be so completely subdued, that it may be grasped across the back and wings, and held breast upwards, while the thighs and underparts are washed. When all the dirt has been removed, wipe off as much of the lather as possible with the sponge, then plunge the bird into the clear water, taking care to rinse out every particle of soap. Next wipe with a soft towel, always stroking the feathers one way, *i.e.*, from head to tail, until as much as possible of the moisture has been removed, when it may be placed near the fire to dry, either in an ordinary exhibition hamper, lined with thin canvas, or in a cage made specially for the purpose, the back and sides of which are composed of wood or canvas, and the top and front of wire. The cage must not be so near the fire as to permit its inmates to be scorched; and if the operator is not pressed for time, he should hold the bird before the fire, with each wing in turn spread open, so as to dry the under side. Whilst drying, they must be watched, and from time to time stroked down, or the feathers may not dry straight and in their proper places.

It is not often that we find Bantams faint whilst being washed, though they do so occasionally; but a little cold water poured over the head, and a teaspoonful of water, with a little whisky or brandy, poured down the throat, quickly revives them, and after a few minutes the work may be continued.

If possible, the washing should be done two or three days previous to the show, and the birds should afterwards be shielded from extreme cold, or draughts, lest they should catch cold, as the washing renders them more susceptible to atmospheric changes.

DUBBING GAME BANTAMS.

Game Bantam cocks only, require to be dubbed. It has been suggested that the Indian Game Bantams, also, should be dubbed, as they frequently are in the large variety; but as their combs rarely grow large enough to be cumbrous, and are, moreover, an ornament, we do not consider it advisable to begin the practice of removing them.

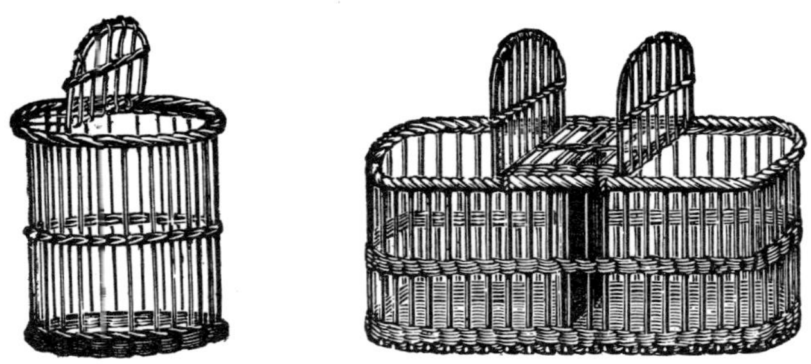

Upright Show Basket. Upright Double Show Basket.
(Mr. Wm. Calway, Sharpness.)

Dubbing is considered by the majority to be a very difficult operation, and many say that it is cruel; but with the latter we do not agree, *if the operation is properly performed.* This, however, we fear, is not always the case, and that the desire to make the head as fine and lean as possible, leads to more suffering than should be permitted. In defence of the practice, we would urge the exceeding pugnacity of all Game cocks, especially during the breeding season; and how much greater are the sufferings of an undubbed bird during and after a battle, than of one whose comb and wattles have been removed.

In order to dub a Bantam successfully, the operator must have a firm, unflinching nerve, and steady hand; also, a pair of sharp, small scissors—strong nail scissors, or those with curved blades, now made specially for the purpose, are the best. A basin of cold water, with a small piece of sponge, are useful aids.

The bird should first have his shanks bound together with a soft wrap, then he should be held by one person in both hands, the third and fourth fingers on each side of the thighs, the second fingers under the breast, while the first fingers and thumbs hold the wings and shoulders in place. The operator then proceeds to cut off the ear lobes and wattles, holding the head securely with his left hand. One clean cut should be sufficient for each, and he must be careful not to cut too deep, because should he open an artery the bird will speedily succumb. Holding the head very firmly, he next removes the comb, cutting from back to front. Two little snips, one on each side over the nostrils, should make all smooth and neat, and complete the work. The head should then be sponged with cold water, and the bird turned down in his usual run.

On the next day, the application of a little vaseline, or pure lard, will be beneficial; and in about ten days the scabs will fall off, when vaseline may again be used; and as soon as the skin has regained its bright red hue, he will be fit to show. Cockerels should not be dubbed before they are six months old; neither is it advisable to delay the operation much beyond that age.

SENDING TO THE SHOWS.

Preparing the birds is not all that is necessary for exhibiting Bantams. Something suitable must be provided in which they may travel securely, and for this purpose we prefer open wicker hampers, lined and topped with stout, but not too closely woven canvas, as this protects them from draughts, while at the same time affording ample ventilation. These hampers may be obtained from Mr. William Calway, and others, for very reasonable prices, and of various sizes, shapes and colours.

An oblong hamper, 3 ft. long by 18 in. wide, and 1 ft. 9 in. deep, with a wicker division down the centre, may be sub-divided with canvas, and thus made to hold four separate pairs of Bantams. This is far more convenient, for conveying several birds, than having a number of small baskets.

Bantams are sometimes shown in pairs, sometimes as single birds, and they must be carefully packed accordingly. By this, we mean, that if they are to be shown singly, each bird must travel separately, having its own address label securely fastened above its compartment. The violation of this rule frequently results in the birds being penned wrongly, and sometimes in their being missed during the re-packing for the return journey, and is thus the cause of both annoyance and additional expense to the exhibitor.

An address label, on which is printed, or written, the class and pen number, is supplied by the secretary of the show, for each exhibit. These generally arrive at least three or four days previous to that fixed for the event; and should this not be the case, the secretary should immediately be communicated with, as it is never wise to despatch birds without their proper labels.

The bottom of the hamper should be covered to the depth of two inches with chaff, or soft, short hay; and if the journey to be taken is a long one, a small lump of meal-dough, a slice of swede, or piece of bread, should be put in also. After being well fed, the birds should have their final polish with the soft handkerchief, then be placed in their hampers, and sent off without delay, the railway carriage being paid through to their destination.

TREATMENT ON RETURN FROM SHOWS.

On their return from shows, the Bantams should again be placed in their cages, and fed with as little delay as possible. Warm soft food should be the first meal, such as bread and milk, Spratt's Poultry Meal, or Liverine. Whichever is used must be given discreetly, and the birds should be watched while they eat, as some are so exhausted by the long fast and journey, following the excitement of the show, that they need coaxing; while others will gorge to an injurious extent; others, again, caring for nothing but to drink water. Each bird should be allowed a sufficient quantity of both food and water, and after two or three hours' rest, have another feed, when a little grain may be given; but on no account must hard grain of any kind be given at first, as they cannot digest it, and thus become crop-bound, and frequently die. If they appear much exhausted, and inclined to mope, one or two "revivers" will quickly set them right again.

When the Bantams will not be required for another show for some weeks, the hens and pullets, and an old cock, may be returned to their usual runs the next evening; and it is best to place them on the perch after the others have gone to roost. But cockerels should each be confined in separate runs, lest they should fight and permanently injure themselves.

CHAPTER XIX.

DISEASES OF BANTAMS, AND TREATMENT WHEN ILL.

ALL Bantams are subject to disease, though some varieties are much more so than others, and all are not equally susceptible to the same form of disease.

The causes of disease are as varied as the trouble itself; but the most fruitful are overcrowding, want of cleanliness in houses, appliances, etc., and injudicious feeding. By careful attention to these sanitary details, which need not be further enlarged upon here, much may be done to prevent its appearance, or, when it does show itself, to arrest its progress.

Some forms of disease are hereditary—that is, transmitted from parent to chick—thus all birds which are not perfectly healthy should be rejected as stock birds. Other kinds are either infectious, or contagious; almost all are more or less so. It is a well-known fact that a bird may come in contact with disease, which is not developed in itself, and yet may give the infection to others in the same run with it; but that, if kept on hygienic principles, and when the general health and condition are good, they are far less liable to it.

By carefully noticing the habits, movements, and appearance of the birds when in health, any illness may be quickly detected, and, upon examination, its nature ascertained; and to aid our readers in this, and in the proper treatment of the sufferer, is the object of this chapter.

As isolation is in most cases necessary, a cage must be provided for the purpose, and used only as a hospital. It should be composed of wood, with a wire front, and be of a convenient size to be moved about; one from 15 to 18 inches square would be large enough. After each patient has been discharged, the cage should be thoroughly disinfected, using carbolic acid, or one of the many preparations now constantly advertised, and lime-washing the whole. By making a practice of putting the cage away in a clean, wholesome state, it will always be ready when, at a moment's notice, it may be required.

During use, the hospital cage should be kept in a moderately warm temperature—about 60 degrees Fahrenheit—and the floor covered with sand, or sawdust, and chaff. Earthenware vessels should be used for food and water, sharp grit should always be within reach, and unless specially forbidden, a small quantity of fresh green food should be given daily.

For an ailing fowl, it is best to vary the diet, as it must consist exclusively of soft food, until the digestive powers permit of a little wheat, canary or hemp seed. A change of green food is also beneficial, say, a grass sod one day, a few dandelion leaves the next, then lettuce, watercress, mustard and cress, etc.; but never give a cabbage leaf, if any of the others mentioned can possibly be procured.

When a bird is convalescent, it must not be returned to its usual home too suddenly, but be gradually accustomed to the change, at first being allowed only an hour's exercise in the open air, and finally being turned out on a fine, mild day.

We will now proceed to describe some of the most common diseases of Bantams, and the most effectual remedies for them.

SIMPLE CATARRH, OR COLD.

This, in Bantams, as in human beings, is the most common ailment, and when promptly treated, is easy to cure. It is caused by sudden changes of temperature, exposure to wet or cold, badly ventilated or damp roosting places, etc., and is chiefly indicated by a thin, watery discharge running from the eyes and nostrils, accompanied by frequent sneezing. When discovered at this stage, the bird should be removed to a warm, airy room, and have three or four doses of camphor, allowing an interval of an hour between each dose. Spirit of Camphor is what we generally use, and one drop is sufficient for a Bantam. It may be administered in the form of a pill, by dropping it on a small piece of bread, or meal dough. If this does not afford relief after a few hours, tincture of aconite is the next remedy, ten drops of which may be given in a quarter-pint of water, and placed where the bird may have frequent access to it. After two days, this must be discontinued, tincture of arsenicum being used instead, in the same proportions, which will usually speedily effect a cure.

During cold and stormy weather, colds may be prevented by placing in the water vessel a small lump of camphor (which any chemist will supply for a penny) in a piece of muslin, weighted to keep it at the bottom. This may be given to all the fowls with advantage. Also when large numbers are suffering from Catarrh, it is not always practicable to take them within doors, and they may be treated with arsenicum in their drinking water, for two or three weeks, if needful; but aconite should never be used excepting when the patients may be kept warm, and free from chills.

Mr. J. H. D. Jenkinson, and others, have prepared remedies for this, which are easily administered, and have also proved invaluable, whose medicines should be given according to the directions which accompany them.

ROUP, AND DIPHTHERITIC ROUP.

This complaint is generally the result of a neglected cold, and also denotes derangement of the digestive system, and, frequently, scrofula. It makes sad havoc, being highly infectious, and often proving fatal, unless skilfully treated.

The symptoms are foul breath, thick, offensive discharge from the nostrils, sneezing, or cough, and inflamed or swollen head round the eyes, a cheesy substance often forming below the eyelid, which, when ripe, must be cut out.

The infected birds should immediately be separated from all others, and kept warm, each bird having its head, and especially the nostrils, bathed with warm water, in which are a few drops of Jeyes' or Condy's Fluid.

Each should then have a pinch of Epsom salts, and be treated with arsenicum, as recommended for Catarrh. If preferred (and in some cases it may be advisable), the medicine may be added to the water used for mixing the soft food, calculating that each bird shall have two drops of the tincture three times a day.

When the mouth and throat are affected, being covered, more or less, with little spots of yellow matter, accompanied by a thick, slimy froth, which causes the sufferer to have great difficulty in breathing, it is called "Diphtheritic Roup." The mouth should be thoroughly cleansed, using either a strong solution of common salt and water, or Condy's Fluid. A camel-hair brush will, perhaps, be the best for reaching down the throat. A lotion, composed of one part perchloride of iron to seven parts of pure glycerine, is useful for painting the affected parts, and destroying the growth, which may soon afterwards easily be removed.

The homœopathic tincture of spongia is the best remedy for internal use, and should be given in the usual manner, *i.e.*, ten drops to a quarter-pint of drinking water.

The bathing must be continued twice daily, until the discharge has ceased to flow, or the cankerous growth is quite removed; and the diet must consist entirely of soft food, and fresh greens, until the patient has made good progress towards convalescence.

For either Roup, or Diphtheritic Roup, when not completely cured by the prescribed remedies, a little flowers of sulphur added to the soft food, once daily, is very beneficial; and for a few days after the other treatment has been discontinued, a tonic should be given, either quinine and cod liver oil capsules, Walton's Tonic Paste, or Hutton's Balm.

DIARRHŒA.

This is caused by sudden changes of temperature, sour, or improper food, by the presence of some irritant in the intestines, or the want of lime, or gravel.

If not excessive, it should not be checked, as it is Nature's mode of relieving the patient; but if prolonged, or accompanied by much distress, an endeavour should be made to discover the cause, and the patient be treated accordingly. If the result of cold, or bad food, arsenicum should relieve, given in one-drop doses, in a little *boiled* milk and bread, every two to four hours, according to the virulence of the attack. Powdered chalk, also, may be added to the soft food; water and green food should be withheld, and the patient be kept warm and dry.

CONGESTION OF THE LIVER.

Congestion of the liver is generally caused by over-feeding, feeding with bad grain, or food of a too stimulating nature, impure drinking water, damp or badly ventilated houses, want of exercise, or a severe shock to the system caused by a sudden chill, and want of grit, or other material, for grinding the food in the gizzard.

The symptoms of the disease are a miserable, moping appearance, the plumage being loose and ruffled, accompanied by loss of flesh, the face becoming pale (and in bad cases, the comb black), and a disinclination for food, and frequently diarrhœa.

Remove the patient to the hospital cage, keep it warm, and administer half a teaspoonful of castor oil, or a good pinch of Epsom salts. The Epsom salts may be repeated each morning for three or four days, food of a light, simple kind being given sparingly, not forgetting fresh green food, of which dandelion, watercress, and a little chopped onion are best, and a liberal supply of sharp grit. Try to discover the cause of the disease, as this will assist one in the selection of remedies, and also to avoid a recurrence of the attack.

Tincture of podophyllum will generally give relief in a few days, if all the exciting causes are guarded against; and bryonia is also very useful.

INFLAMMATION.

The bowels, and other internal organs of Bantams, sometimes become inflamed, either as the result of cold, improper feeding, or injury.

Not much can be done for the sufferer beyond keeping it warm and quiet, feeding it sparingly on soft food, and giving it aconite. Ten drops of the latter in a quarter-pint of water is the proportion, and the bird should be allowed to drink freely every two hours until the inflammatory symptoms subside. The food and water should not be given cold, neither should they be left within reach at all times. After convalescence, a "tonic," or "reviver," should be given for a few days.

CONSUMPTION.

This is commonly called "wasting," or "going light." The bird loses flesh, becomes pale and sunken in face, and has a general moping and ruffled appearance, accompanied in most cases by loss of appetite.

Young birds are the most frequent victims, and if the disease has made much progress before treatment is commenced, we believe a cure is impossible. If, however, the symptoms are noticed early, the disease may be arrested, and stamped out by strict attention to the sanitary arrangements, and by giving the most nourishing food, "a little at a time, and often." A cod liver oil capsule should be given every day, and the drinking water coloured with tincture of iron.

This is not a case for the hospital cage, but one in which the patient should have unlimited exercise in a warm, dry, sunny atmosphere.

SCROFULA.

This is an exceedingly troublesome disease, and consists of minute particles of matter intermixed with the blood. The disease may lie dormant and unsuspected for a considerable length of time, until it is excited into activity by cold, or some accident, when it will develop into roup, abscesses, ulcers and sores, liver disease, consumption, etc. The corpuscles congregate, and find a weak place, where they may do their work of destruction.

Scrofula is hereditary, and birds with this taint in their blood are far more liable to disease of all kinds than others. It is also produced by overcrowding in ill-ventilated houses, bad food, and such things as tend to injure the health generally. Birds that are known to be thus affected should not be used for breeding from, and the greatest care should be exercised with regard to their proper housing and feeding; and if small doses of prepared charcoal, iron, sulphur, or phosphorus be given, they will be found of benefit. If, however, this disease should develop in an acute form in a valuable specimen, it would be the wisest plan to send it for treatment to Mr. J. H. D. Jenkinson, who has made a special study of such ailments for many years.

CHOLERA.

This highly contagious and much dreaded disease is the cause of great havoc in the poultry yard, when once it makes its appearance there, as but little can be done for the relief of the sufferers; but, fortunately, it is of rare occurrence in England, when the birds are kept under proper sanitary conditions. The symptoms are excessive thirst, accompanied by diarrhœa, the excrements being of a greenish colour at first, and afterwards thin and white. This is followed by extreme weakness, emaciation, inability to stand, and cramp—then death. The best plan is instantly to kill the affected birds, and burn the bodies; thoroughly disinfect all the runs and houses; and give to the rest of the Bantams ten drops of strong tincture of camphor in half a pint of drinking water, for a few days, as a preventive.

CRAMP.

This does not attack adults so frequently as chickens, and is usually the result of their being kept in cold, damp localities, or where they have not sufficient exercise.

It is an affection of the feet and legs, and will be discovered by the manner in which the chickens walk, and afterwards by the toes contracting, or curling up. The sufferers should immediately be removed to a warm, dry run, where they may have as much exercise as possible; and be fed on a more nourishing diet.

Bathing the feet with warm water, and afterwards well rubbing them with turpentine, or St. Jacob's Oil, is a valuable aid to their recovery.

LEG WEAKNESS.

Growing chickens sometimes suffer from this cause, when insufficiently supplied with bone-forming materials in their food. Instead of walking or running about in an easy and sprightly manner, they are found squatting on the ground, or shuffling along on the hock joint, assisted by the wings.

The remedy, of course, is to change the diet, using oatmeal, sharps, and bone-meal, or Liverine, as the chief food. A cod liver oil capsule may be given daily, and Parrish's Chemical Food added to the drinking water.

Rubbing the legs and feet with Elliman's Embrocation will facilitate the circulation, and thus aid recovery.

CROP-BOUND.

This trouble is the result of either injudicious feeding, or of the bird having swallowed some hard substance which obstructs the outlet of the crop, and causes the food to accumulate, and form a hard lump, the crop being distended much more than is usual after a hearty meal.

If the mischief is discovered early, and is only the result of improper feeding, it can generally be quickly relieved by pouring about half a teacupful of warm water, to which has been added a large spoonful of treacle, down the throat, and then gently working the crop with the fingers. This will probably cause the patient to vomit, and thus relieve it of some of the inconvenience, and the remainder will pass through the other digestive organs.

Should this plan fail, it may be necessary to open the crop to remove the obstruction. Strip some of the feathers off the upper part of the breast, and with a sharp penknife, or pair of scissors, make a lengthwise incision, about an inch long. The mass of food can then easily be removed by means of a small spoon. When the crop is emptied, it is advisable to pass the finger inside, well oiling it first, to feel if any foreign matter remains, and thoroughly rinse the cavity with warm water and a little Condy's Fluid.

The hole must then be sewn up, using a fine needle and white silk for the purpose. Each skin must be sewn separately, and two or three stitches will be necessary for each. It is best to tie the thread for each stitch, and cut it short; and the inner skin, of course, must be done first.

The sufferer must be fed sparingly on bread soaked in milk, Spratt's Food, and meal dough, until the wound is quite healed up; and no water must be given for at least two days.

EGG-BOUND.

It is not at all an uncommon occurrence for a hen to be egg-bound, especially during cold weather early in the season. She will be noticed to frequently visit her nest, and from time to time leave it again without depositing her egg. She will also appear much distressed, her tail and wings drooping.

The principal causes of this are either the unusually large size of the egg, or contraction of the egg passage.

Bathe the abdomen and vent with warm water for a few minutes, then well grease the vent with pure lard or olive oil, and place the bird in a covered basket, in the bottom of which is a thick bed of hay, and place her in a warm, quiet room.

If after an hour's time she has not laid, she should again be bathed, or, better still, steamed, by holding the vent over a jug of very hot water, again smearing the part with oil. In obstinate cases, a spoonful of warm treacle, or a very small dose of castor-oil may be administered. After the egg has been laid, bathing with a weak solution of arnica is beneficial; and the hen should be kept quiet for a few hours, well fed, and then returned to her run.

GAPES.

This is the most common and most troublesome of chickens' ailments, and it attacks only young birds from two to about eight weeks old, except in rare instances. Beyond this age they are seldom troubled, and adult fowls never succumb to it.

Gaping is not the disease, but the result of it, and the indication that disease exists in the chicken so afflicted. The poor little thing stands with all its feathers ruffled, and wings drooping, and constantly opening and shutting its beak. The more frequently it gasps, the more urgent is its need of help.

The cause of this distress is a small, thread-like worm of a reddish-white colour, the origin of which is still a matter of conjecture; but its effects upon our feathered pets are well known. These parasites attach themselves to the mucous membrane lining the windpipe, and from two to a dozen may be found in one victim, and, if they are not removed, cause death by suffocation.

Where only a few chickens are attacked, they may be treated individually, and the worms be removed by means of a small quill feather, which has been stripped to within half an inch at the point, and dipped in turpentine. Shake off all that does not adhere, insert it down the windpipe, and give it a sharp twist round, when, if properly done, it should dislodge the worms, and bring them up.

When whole broods are attacked, as is frequently the case, the easiest plan is to destroy the gape-worms by inhalation, or fumigation. "Zoagrion," "Kalydé," and "Camlin" are, perhaps, the best preparations to use for this purpose. The chickens are placed in a box, and the powder is blown into it by means of a distributer, specially made for the purpose. Usually, one dose is sufficient to destroy the worms.

Another mode is to place the box or basket containing the chickens over the fumes of burning carbolic acid. Great care is needed in using this remedy, or the chickens, as well as the gape-worms, may be suffocated.

After an attack of the Gapes, the chickens are often very weak, and will consequently require a little extra care and attention until they have recovered their usual strength and appetite.

As preventive measures, "Anti-Vermine" is a good preparation; also, lime-water in place of the ordinary drinking water; and over-crowding, tainted ground, and uncleanliness of every kind must be avoided.

WORMS AND LICE.

Intestinal worms sometimes prove very troublesome to Bantams, and cause them to become thin and out of condition generally.

If their presence is suspected, the droppings should be examined, when, if they exist, some will be found amongst them.

The homœopathic tincture of cina is a sure remedy, and should be given in the drinking water, and continued until the bird has quite recovered.

Lice are often found amongst the feathers of Bantams, and cause them much discomfort, besides injuring the plumage.

There are two or three very common varieties, the large grey insect, which is easily discernible, being almost the size of a grain of canary seed. Another kind is exceedingly small, red or brown in colour, and very quick in its movements.

When young chickens are infested with these parasites they cannot thrive, but mope and chirp miserably; and, if examined, their heads and necks will be found almost destitute of feathers, and these parts, and under the wings, covered with lice. They should be dusted over with pyrethrum powder, which will, if it is good, speedily destroy them. Another plan is to touch the head and the nape of the neck with an ointment made of the following: Mercurial ointment and pure lard, of each 1 oz.; flowers of sulphur and crude petroleum, of each ½ oz. Mix all thoroughly together, and apply. This will keep good for some months, if in a covered jar, in a cool place.

Adult fowls, also, if not provided with a dust-bath, are troubled in the same way, and the feathers are destroyed, and lose their lustre. The most effectual remedy is to wash the bird thoroughly, using a good dog soap—of course, well rinsing it in clean water.

A dust-bath should always be provided for Bantams of all ages. It should be kept dry, and frequently replenished with dry ashes, mixed with flowers of sulphur. Wood ashes, if obtainable, are excellent for the purpose; but sifted coal ashes, or sand, or even road scrapings, will do, and thus prevent much trouble and annoyance.

SCALY LEGS, OR ELEPHANTIASIS.

The feather-footed varieties of Bantams are the most frequently troubled with scaly legs, though the disease sometimes makes its appearance on the other varieties. It is caused by a parasite lodged under the scales, and if not promptly treated, results in the growth of a rough, unsightly scurf, which extends from the hock to the toes.

The legs should be thoroughly soaked in warm water for a few minutes, and then brushed with a good nail-brush and carbolic soap, after which sulphur ointment should be well rubbed into the affected parts. A little paraffin is also a useful application, and the Liverine Company has lately brought out an ointment called "Unction," which is an excellent remedy. The washing and rubbing should be repeated daily, until the patient is cured, taking care, however, not to scrub so hard as to make the legs bleed. Flowers of sulphur should be added to the soft food, and a more liberal supply of green food be given, if possible, until the disease is cured.

As Elephantiasis is very contagious, the bird should be isolated; or, if more than one is affected, the treatment should extend to all; the houses, perches, etc., being frequently lime-washed, with a little paraffin, or carbolic acid, added to the lime.

FEATHER CURL.

This is caused, it is supposed, by parasites. The feathers, usually on the breast and thighs, become slightly twisted and ruffled; at first, only a few feathers being affected, and that so slightly as to be almost unnoticed. In a few days it becomes worse, spreading over a larger area, and the feathers curling back, almost like those of a Frizzled Bantam. Sometimes the feathers on the back are affected, and all are liable to it

This is one of the hereditary complaints, but has been known to make its appearance several times where heredity could not account for it, and this leads us to suppose that it may be slightly contagious. It is also more prevalent during a dry season than a wet one.

A perfect cure has never yet been discovered for feather curl, and Bantams suffering from it are quite spoiled for the show pen.

WHITE COMB.

This is a disease from which the large-combed varieties of Bantams sometimes suffer, but, if promptly attacked, soon yields, and need cause little anxiety.

The chief causes are over-crowding in damp, ill-ventilated roosting places, improper feeding, and lack of green food.

The comb first appears to be covered with a white dust, which gradually becomes scales, or scurf, extending from the comb over the head and neck, and causing loss of the feathers.

The patient should immediately be removed to dry, airy quarters, and have the affected parts carefully washed in warm water (using a very little good soap), and afterwards dried on a soft cloth. When dry, rub well with sulphur ointment, or pure glycerine and carbolic acid, in the proportion of 3 drops of acid to a tablespoonful of glycerine. Olive oil may be used instead of glycerine. This application should be used daily, until a cure is effected; and either tincture of sulphur used in the drinking water, or flowers of sulphur added to the soft food. Green food should be given liberally.

After convalescence, a course of tonics will be needed, such as iron, or cod liver oil and quinine capsules.

FROSTED COMBS.

During very severe weather, the large combed Bantams often suffer from frost bitten combs, which, if not quickly attended to, cause great disfigurement, if not actual loss of life.

The first symptoms are blackening of the spikes, or, if a bad case, of the whole comb; and if the latter, the case is a serious one. The circulation of the blood must be restored by friction; rubbing with snow, or cold water, at first, while supporting the comb with the palm of one hand and rubbing with the fingers of the other. Afterwards, a liniment composed of equal parts of olive oil, whisky, camphor, and laudanum, has been found to answer admirably, and should be rubbed in briskly, two or three times a day, until the natural colour is quite restored.

The bird should have a "reviver" given to him on alternate nights, for a short time; and should any symptoms of Catarrh, or Liver Complaint appear, which may be brought on by the shock to the system, he must be promptly treated accordingly.

WOUNDS AND BRUISES.

Bantams frequently meet with accidents of one kind or another, fighting being the cause of the greatest number of wounds. If the wound is bleeding profusely, it should be bathed with cold water to check the flow. If dirty, and bleeding very little, bathe with luke-warm water until quite clean; then, with a pair of sharp scissors, remove any jagged bits of skin, and, if necessary, draw the skin together, using a fine needle, and either fine white cotton or silk, and tying each stitch separately. When this has been done, apply a little calendula, diluted with water, which will help to form a scab, and heal the wound. Afterwards, pure lard or vaseline may be used.

Tincture of calendula may be used with water for bathing any bruises when the skin is broken, and we always find it a valuable remedy in such cases.

EGG-EATING AND FEATHER-EATING.

These can scarcely be called *diseases* of Bantams, but they certainly are *bad habits* of rather common occurrence, particularly amongst birds kept in a confined run.

The cause is, in both cases, the same—the lack of something which they need, and this should immediately be supplied. In the case of egg-eating, it will most probably be shell forming materials that are required, and old mortar and crushed oyster shells should accordingly be given. If the habit is still continued (and when it has once been formed it is difficult to cure), an egg shell may be filled with a paste made exceedingly hot with mustard or cayenne, and laid in the nest. If this fails, the culprit should be killed, unless a valuable one, and in the latter case, the only remedy would be to procure a nest box so constructed that the egg, as soon as deposited, rolls down an inclined board into a place of safety beneath.

Feather-eating is very troublesome, as it quite spoils the appearance of the birds. Again, it is generally the hens that are in fault, and they will sometimes quite strip the neck, and part of the breast, of the cock, or other hens.

If they have been kept in a confined run, they should, if possible, be allowed a free range in a meadow, where they will find plenty of occupation in hunting for insects; and if this plan cannot be adopted, the birds must have a liberal supply of all necessaries given to them, including abundance of fresh green food, which would be best suspended in bunches. When the latter is difficult to obtain, swedes, mangold wurzels, apples, apple parings, and even raw potatoes cut in halves, will serve to keep them out of mischief, and also prove of value in other ways.

If these measures fail, as a last resort the culprit must be muzzled by means of a piece of fine copper wire passed through the nostrils, and round the upper part of the beak, and fastened at the side. This, if properly done, prevents the bird closing its beak tightly enough to eat feathers, though it does not prevent its being able to take food properly, and will thus answer the purpose desired.

ADMINISTRATION OF MEDICINES.

In conclusion, we will only add that in all the instances where we have prescribed homœopathic tinctures (aconite, arsenicum, bryonia, cina, podophyllum, spongia, and sulphur), the dose, unless otherwise directed, is in the proportion of ten drops to a quarter-pint of drinking water, to be given fresh every day, and in a pot—not a metal vessel—and left where the bird may drink at will. We have had considerable experience in the use of these, and always found them reliable; but we do not confine ourselves only to homœopathy, and have found the pills, etc., prepared by Messrs. Jenkinson, King, Walton, and others, of great value.

CHAPTER XX.

THE BANTAM CLUB.

EARLY in the year 1880, some letters were published in "The Fanciers' Chronicle," about Brown-red Game Bantams, the object being to excite a greater interest in this variety, and procure separate classes for them at the Poultry Exhibitions, where they generally had to compete against Black-reds, if not against all colours of Game Bantams. An anonymous writer proposed the formation of a "Brown-red Game Bantam Club," but this proposal did not meet with much favour, and in a reply from Messrs. T. H. and A. Stretch, published the following week, they pointed out that it would be much better to form a club to include all the varieties of Bantams, and to be called "The Bantam Club," as it would be better supported, and would thus have greater influence with show committees, and benefit the whole of the Bantam fanciers.

The correspondence lasted several months, during which the interesting fact was disclosed that there were at that time, in the British Isles, over 1,500 Bantam exhibitors who would share in the advantages gained by the club; but it was not until February, 1881, that a meeting was called to discuss the subject. This was held at the Charing Cross Hotel, London, and was well attended, Mr. Phelps, of Ross, taking the chair, and in a few sensible remarks, urging the need for a club, as at the majority of shows the numerous varieties of Bantams were restricted to one or two classes; and showing that the club could do much to remedy this state of things, and obtain a better classification for, and a recognition of, the different breeds. The matter was then freely discussed, and the following proposition made:—"That a club be formed, to be called The Bantam Club, for the purpose of promoting the better classification of the various breeds of Bantams at shows." A list of the names of about twenty-five ladies and gentlemen, who had expressed their willingness to join, was read over; and it was decided that a provisional committee, consisting of Messrs. Wright, Barnes and Oscroft, should issue a circular to the Bantam Fancy, inviting persons to join, and give their advice, and that as soon as the general feeling had been discovered, a meeting should be held, the rules drawn up, and officials appointed.

Subsequently, the rules were drawn up, and the first officials appointed were:—Mr. Phelps, president; Mr. Oscroft, treasurer; and Mr. E. Wright, secretary. The committee consisted of Messrs. H. B. Astley, J. Barnes, O. E. Cresswell, J. E. Gunn, G. Hall, J. A. Hewetson, T. H. Stretch, and A. Stretch; and the number of members during the first year was thirty-six.

The annual subscription was, in the beginning, one guinea each, but in 1882, at the general meeting, held in the Crystal Palace during the show week, it was reduced to ten shillings, the committee thinking that this would induce more members to join; and so it did, the number in 1883 having increased to sixty.

From time to time, at the general meetings, it was found advisable to make some alteration in, or addition to, the rules, until in 1884 they assumed their present form; and as they fully explain the principles and mode of procedure of the club, we will give them in full.

RULES.

1.—That the Club be called "THE BANTAM CLUB."

2.—That the objects of the Club be—

(*a*.) To obtain better classification of the several varieties of Bantams at Shows; and for this purpose to guarantee to the committee of any show the deficiency, or any part thereof (if any), between the amount of the entrance fees and prizes awarded in any classes for Bantams, or to offer special prizes for the same.

(*b.*) To hold a Show of Bantams (annually) open to members of the Club only, or to exhibitors generally, as the committee may decide.

(*c.*) The advancement and protection of the interests of exhibitors and breeders of Bantams.

(*d.*) The suppression of fraud and dishonourable conduct in all matters connected therewith; and

(*e.*) Generally to promote the improvement and exhibition of the Bantam varieties.

3.—That the Club consist of members whose annual subscription shall be 10s., payable in advance on the 1st of January in each year, and that any member whose subscription shall be in arrear on 1st of August in each year shall have notice of such sent him by the honorary secretary, and he shall not be allowed to make entries at the reduced fee at shows supported by The Bantam Club until his subscription shall have been paid, and if not paid before 31st December in any year, he shall be considered to have ceased Membership.

4.—That the officers of the Club shall consist of a president, vice-president, two auditors, secretary, and treasurer.

5.—That the committee shall consist of the officers of the Club, who shall be *ex-officio* members thereof, and twelve other members, who shall manage the business of the Club. Five of the committee (one, at least, of whom shall be an officer) shall form a quorum. Five members of the general committee, together with the president and secretary, shall be elected annually by the general committee an executive committee to decide grants to Shows, guarantees for and arrangements of classes, the secretary to obtain the sanction of the members of such committee before arranging with any Show.

6.—That the officers and members of the committee be elected annually at the general meeting held during the Crystal Palace Show, their term of office to commence the 1st of January following. Half of the committee shall retire annually, but shall be eligible for re-election.

7.—That the election of officers and committee shall be conducted by means of voting papers; all nominations to be sent in on or before October 1st in each year; voting papers with such nominations to be sent out to all members whose subscriptions are not in arrear by October 7th. Such voting papers, properly filled in and signed, must be returned to the secretary at or before the annual meeting during the Crystal Palace Show, when they will be examined by scrutineers appointed there, and the result declared.

8.—That a general annual meeting of the members of the Club shall be held at the Crystal Palace during the Show week, or at such other time and place as the committee shall appoint. That seven of the committee may at any time, in writing, convene a special general meeting for the discussion of any matter connected with or affecting the interests of the Club, notice of such general annual or special general meeting shall be sent by the secretary to the officers and members, fourteen days, at least, before such meeting.

9.—That all members be elected by the committee, and that the committee be empowered to elect as honorary members of the Club, without subscriptions, ladies and gentlemen distinguished as fanciers or judges of Bantams.

10.—That candidates for election send their names and addresses to the secretary, who shall, seven days at least before the day of election, send a list of the names and addresses of such candidates to the officers and members of the committee.

11.—That if any member be proved, to the satisfaction of the committee, to have acted dishonourably, he shall be suspended from being a member for such term as the committee may adjudge, or be expelled, and shall be ineligible for re-election. That if any member who shall be suspended or expelled be dissatisfied with the decision of the committee, and shall, within twenty-one days from such suspension or expulsion, give notice thereof, in writing, to the secretary of his intentions so to do, he may appeal against such decision at a general annual or special general meeting duly convened.

12.—That exhibitors and others be requested to bring to the notice of the committee any offences against the objects of the Club; and that the committee be empowered, if they see fit, to do all things necessary to prosecute with effect the offenders.

13.—That any member who shall exhibit any bird which shall be proved, to the satisfaction of the committee, to have been trimmed, drawn, or otherwise tampered with, he shall be suspended for such term as the committee may adjudge, or be expelled from the Club, and be ineligible for re-election.

14.—That members be requested not to exhibit at any Poultry Show which does not publish with its schedule the names of the judges, and as far as possible the classes upon which they shall adjudicate, any unavoidable changes to be previously notified in the poultry paper (if possible).

15.—That a list of the officers, committee, and members, together with a balance sheet and report, be sent annually to the members, and to any other person applying for the same.

16.—That no alteration shall be made in these rules, except at the general annual meeting, or at a special general meeting (duly convened). Notice of any proposed alteration must be sent to the secretary fourteen days, at least, before the meeting; and the secretary shall give seven days' notice of such meeting, and of the proposed alteration, to the officers and members of the Club.

The subscriptions, with the exception of a small sum necessary for printing, stationery, postage, etc., were all used for providing cups and medals, and in assisting different show committees to offer a better classification for Bantams.

The first show held under The Bantam Club rules was at Stratford-on-Avon, in the Autumn of 1881, and here two classes were provided for Game Bantams, one for Rosecombs, one for Sebrights, and one for any other variety. The Club also assisted four other shows during the same year, and at the end had a balance in the bank of £18 towards the next season.

The Club prospered, the number of members rapidly increased, and nearly all the most influential breeders and exhibitors joined it, and thus enabled it to succeed in its object. The annual Club Show was held in connection with various other shows, the first being in 1882 at Kendal, the next at Cheltenham, then the Dairy Show, Islington; also at Manchester, and several times at the Crystal Palace. Of course, at these shows, there was a magnificent classification provided, and it is scarcely necessary to say that they were well supported by the members of the Club.

Latterly, there has been a decrease in the number of members, but in the year 1892, there were still sixty-one, and the officers were as follows:

President:
MR. R. L. GARNET.

Vice-President:
MR. F. BROOKE.

Auditors:
MR. J. GUNN and MR. A. STRETCH.

Honorary Secretary and Treasurer:
REV. F. COOPER and MR. J. C. PRESTON.

Committee:

MR. H. AINSCOUGH.	MR. O. E. CRESSWELL.
,, G. FURNESS.	,, T. C. HEATH.
,, F. GEARY.	,, E. WALTON.
,, T. H. STRETCH.	,, W. F. ENTWISLE.
,, C. W. BRIERLEY.	,, E. WRIGHT.
,, H. S. RAINFORTH.	,, W. MOORE.

Executive Committee:

MR. H. AINSCOUGH.	MR. F. BROOKE.
,, E. WALTON.	,, W. F. ENTWISLE.
,, T. H. STRETCH.	,, E. WRIGHT.

Since the annual meeting at the Crystal Palace, in November, 1893, The Bantam Club has ceased to exist.

INDEX.

Printed by Nene Litho, Earls Barton, Northants.
Bound by Weatherby Woolnough, Wellingborough, Northants.